John William Dawson

On the Conditions of the Deposition of Coal

more especially as illustrated by the coal-formation of Nova Scotia and

New Brunswick

John William Dawson

On the Conditions of the Deposition of Coal
more especially as illustrated by the coal-formation of Nova Scotia and New Brunswick

ISBN/EAN: 9783337100698

Printed in Europe, USA, Canada, Australia, Japan

Cover: Foto ©berggeist007 / pixelio.de

More available books at **www.hansebooks.com**

[*From the* QUARTERLY JOURNAL *of the* GEOLOGICAL SOCIETY *for*
May 1866.]

ON THE

CONDITIONS

OF THE

DEPOSITION OF COAL,

MORE ESPECIALLY AS ILLUSTRATED

BY THE

COAL-FORMATION OF NOVA SCOTIA

AND

NEW BRUNSWICK.

BY

J. W. DAWSON, LL.D., F.R.S., F.G.S.,
PRINCIPAL OF M'GILL COLLEGE, MONTREAL.

ON THE

CONDITIONS OF THE DEPOSITION OF COAL,

MORE ESPECIALLY AS ILLUSTRATED

BY THE

COAL-FORMATION OF NOVA SCOTIA

AND

NEW BRUNSWICK.

BY

J. W. DAWSON, LL.D., F.R.S., F.G.S.,
Principal of M'Gill College, Montreal.

[PLATES V.-XIII.]

CONTENTS.

§ I. Introduction.

In several former papers presented to this Society, I have endeavoured to illustrate the arrangement of the Carboniferous rocks of Nova Scotia, and to direct attention to their organic remains, the structures found in their coals, and the evidence which they afford as to the mode of accumulation of that mineral. The present paper is intended as the summing up and completion of these researches, with the addition of the new facts resulting from a careful study of the microscopic structure of more than seventy beds of coal occurring in the South-Joggins section, and of the fossil plants associated with them. These results will, I hope, throw much additional light on some of the more difficult problems connected with the theory of the accumulation of vegetable matter in the Carboniferous period, and its conversion into coal.

The subjects to which I propose to direct attention may be conveniently arranged under the following heads :—

(1) General considerations relating to the physical, conditions of the Carboniferous period in Nova Scotia.

(2) Details of the character and contents of the several beds of coal in the Joggins section, arranged in the order of Logan's Sectional List.

(3) Remarks on the genera of animals and plants whose remains occur in the coal, and on their connexion with its accumulation.

§ II. General Considerations relating to Physical Conditions.

1. *Physical Characters of the several Coal-formations.*—The total vertical thickness of the immense mass of sediment constituting the Carboniferous system in Nova Scotia may be estimated from the fact that Sir W. E. Logan has ascertained by actual measurement at the Joggins a thickness of 14,570 feet ; and this does not include the lowest member of the series, which, if developed and exposed in that locality, would raise the aggregate to at least 16,000 feet. It is certain, however, that the thickness is very variable, and that in some districts particular members of the series are wanting, or are only slenderly developed. Still the section at the Joggins is by no means an exceptional one, since I have been obliged to assign to the Carboniferous deposits of Pictou, on the evidence of the sections exposed in that district, a thickness of about 16,000* feet; and Mr. Brown has estimated the Coal-formation of Cape Breton, exclusive of the Lower Carboniferous, at 10,000 feet in thickness†.

When fully developed, the whole Carboniferous series may be arranged in the following subordinate groups or formations, the limits of which are, however, in most cases not clearly defined :—

a. *The Upper Coal-formation.*—It consists of sandstones, shales, and conglomerates, with a few thin beds of limestone and coal. *Calamites Suckovii, Annularia galioides, Cordaites simplex, Alethopteris nervosa, Pecopteris arborescens, Dadoxylon materiarium, Lepidophloios parvus,* and *Sigillaria scutellata,* are among its characteristic vegetable fossils.

* Quart. Journ. Geol. Soc. vol. i. p. 329.   † Ibid. vol. vi. p. 116.

b. *The Middle Coal-formation, or Coal-measures proper.*—This series includes the productive beds of coal, and is destitute of properly marine limestones. Beds tinged with peroxide of iron are less common in this formation than in any of the others. Dark-coloured shales and grey sandstones prevail, and there are no conglomerates. *Sigillariæ* and *Stigmariæ* of many species are the most conspicuous and abundant fossils, but Ferns, *Cordaites*, and *Calamites* are also extremely abundant, and all the genera of Carboniferous plants are represented. Many beds, especially those in the vicinity of layers of coal, contain minute *Entomostraca*, shells of the genus *Anthracomya* (*Naiadites*), *Spirorbis carbonarius*, and remains of ganoid and placoid fishes.

c. *The "Millstone-grit" Formation.*—This name, though not in all cases lithologically appropriate, has been borrowed from English geology to designate the group of sandstones, shales, and conglomerates, destitute of coal, or nearly so, and with few fossil plants, which underlies the Coal-measures. In its upper and middle part it includes thick beds of coarse grey sandstone holding prostrate trunks of coniferous trees (*Dadoxylon Acadianum*). In its lower part red and comparatively soft beds prevail.

d. *The Lower Carboniferous Marine Formation.*—The essential features of this formation are thick beds of marine limestone, characterized principally by numerous Brachiopods, especially *Productus Cora*, *P. semireticulatus*, *Athyris subtilita*, and *Terebratula sufflata*[*], with other marine invertebrates. Associated with these limestones are beds of gypsum, and they are enclosed in thick deposits of sandstone, clay, and marl, of prevailing red colours.

e. *The Lower Carboniferous Coal-measures, or Lower Coal-measures.*—In some localities these resemble in mineral character the true Coal-measures. In others they present a great thickness of peculiar bituminous and calcareous shales. They usually contain in their lower part thick beds of conglomerate, and coarse sandstone which in some places prevail to the exclusion of the finer beds. The characteristic plants of these beds are *Lepidodendron corrugatum*, and *Cyclopteris Acadica*, with *Dadoxylon antiquius*, and *Alethopteris heterophylla*[†]. They also contain locally great quantities of remains of fishes, and many Entomostracans, among which are *Leaia Leidyi* and an *Estheria*, also *Leperditia subrecta*, Portlock, *Beyrichia colliculus*, Eichw., and a *Cythere*[‡], probably new.

The last two groups are equivalent to the "Sub-carboniferous" of some American geologists; but independently of the objection to the use of a term which would seem to imply a formation under, and distinct from, the Carboniferous, and of undetermined age, I find in Nova Scotia no reason, either palæontological or stratigraphical, for any greater distinction than that implied in the term Lower

[*] See Davidson "On Lower Carboniferous Brachiopoda from Nova Scotia," Quart. Journ. Geol. Soc. vol. xix. p. 158.
[†] Dawson, "On the Lower Coal-measures," &c., Quart. Journ. Geol. Soc. vol. xv. p. 62.
[‡] Prof. Jones has kindly determined these species.

Carboniferous. The Lower Coal-measures are, it is true, more distinct in their flora from the Middle Coal-measures than the latter from the Upper Coal-formation ; but still many species are common to the two former, and the difference is small as compared with that between the Lower Carboniferous and the Upper Devonian. The Devonian rocks are also in this region unconformable to the Carboniferous, having been disturbed and altered prior to the deposition of the latter ; while no want of conformity, except of the local character hereafter to be noticed, occurs in the Carboniferous.

2. *Physical Conditions attending the Deposition of the Coal-formations.*—The conditions of deposit implied in the mineral character of the several formations above described, would appear to be of three leading kinds:—(1) The deposition of coarse sediment in shallow water, with local changes leading to the alternation of clay, sand, and gravel. This predominates at the beginning of the period, recurs after the deposition of the marine limestones in the formation of the "Millstone-grit," and again prevails in the Upper Coal-formation. (2) The growth of corals and shellfish in deep clear water, along with the precipitation of crystalline limestone and gypsum. These conditions occurred during the formation of the Lower Carboniferous Limestone and its associated gypsum. (3) The deposition of fine sediment, and the accumulation of vegetable matter in beds of coal and carbonaceous and bituminous shale, and of mixed vegetable and animal matters in the beds of bituminous limestone and calcareo-bituminous shale. These conditions were those of the Middle Coal-formation.

Within the limits of Nova Scotia, these conditions of deposition applied, not to a wide and uninterrupted space, but to an area limited and traversed by bands of Silurian and Devonian rocks, already partially metamorphosed and elevated above the sea, and along the margins of which igneous action still continued, as evidenced by the beds of trap intercalated in the Lower Carboniferous[*]; while about the close of the Devonian period still more important injections and intrusions of igneous matter had occurred, as shown by the granitic dykes and masses which traverse the Devonian beds, but have not penetrated the Carboniferous[†]. There is evidence, however, in the Carboniferous rocks of the Magdalen Islands and of Newfoundland, and in the fringes of such rocks on parts of the coast of Nova Scotia[‡] and New England, that the area in question was only a part of a far more extensive region of Carboniferous deposition, the greater part of which is still under the waters of the Atlantic and of the Gulf of St. Lawrence.

There is ample proof that most of the coarser matter of the Carboniferous rocks was derived from the neighbouring metamorphic ridges : but much of the finer material was probably drifted from more distant sources. There seems no good reason to doubt that in the Carboniferous period, and especially in those portions of it in

---

[*] Dawson, Quart. Journ. Geol. Soc. vol. i. p. 329.
[†] Dawson, Canadian Naturalist, 1860, p. 142.
[‡] Jukes's 'Newfoundland ;' 'Acad. Geology,' p. 274.

which the areas now under consideration were in the condition of
shallow seas or swampy flats, the greater part of the Laurentian and
Silurian districts of North America existed as land; while the great
number of Coal-formation plants common to Europe and America
may indicate the existence of intermediate lands now submerged.
From such lands, undergoing waste during the long Carboniferous
time, the materials of the shales and finer sandstones may have been
derived.

Taking this view of the source of the sediment, we should infer
that the time of the formation of the marine limestones was that of
greatest depression of the land, when the local ridges of older rock
were mere reefs and islets, and when sediment from more distant
lands was deposited only at intervals. We should also infer that the
time of the formation of the coal-beds was that of greatest elevation,
when the former sea-bottoms had become land-surfaces or flats,
exposed only to occasional inundation, and when rivers were bearing
downward from large continental regions great quantities of fine
silt. Further, the conditions of the Millstone-grit and of the Newer
Coal-formation must have been of an intermediate character, re-
quiring wide sea-areas receiving great quantities of sediment, and
on this account, as well as because of their shallowness, unfavourable
to marine life, while the areas of vegetable growth were also of
limited extent.

It would also follow that when the Lower Coal-measures and
conglomerates were formed, the land was slowly subsiding; that in
the time of the marine limestones it attained to its greatest depression, and long remained nearly stationary; that in the Millstone-
grit period there was re-elevation, and that in the period of the
Middle Coal-formation and Newer Coal-formation there was again
subsidence, slow and interrupted at first, but subsequently of greater
amount. From the absence of Permian deposits it may be inferred
that elevation again took place at the close of the Carboniferous period,
to such an extent as to preclude further deposition in the area in
question; while the red sandstone and trap of Mesozoic age indicate
the recurrence at that time of conditions somewhat similar to those
of the beginning of the Carboniferous period.

The general phenomena of deposition above indicated apply to all
the Carboniferous areas of Nova Scotia and New Brunswick, and, so
far as known, to those of the Magdalen Islands and of Newfoundland.
But, as I have pointed out in 'Acadian Geology,' numerous local
diversities occur, in consequence of the interference of the older
elevated ridges with the regularity of deposition. In some places the
entire Lower Carboniferous series seems to be represented by conglomerates and coarse sandstones. In others, the Lower Coal-
measures, or the Marine Limestones, or both, are extensively developed. These local differences are, on a small scale, of the same
character with those which occur on a large scale in the Northern
and Southern Appalachian districts and Western districts of the United
States, and in the different coal-areas of Great Britain and Ireland,
as compared with each other and with the Carboniferous districts of

America. On the whole, however, it is apparent that certain grand features of similarity can be traced in the distribution of the Carboniferous rocks throughout the northern hemisphere.

It is further to be observed that in Nova Scotia and New Brunswick, as well as in Eastern Canada, disturbances occurred at the close of the Devonian period which have caused the Carboniferous rocks to lie unconformably on those of the former; and that in like manner the Carboniferous period was followed by similar disturbances, which have thrown the Carboniferous beds into synclinal and anticlinal bends, often very abrupt, before the deposition of the Triassic Red Sandstones. These disturbances were of a different character from the oscillations of level which occurred within the Carboniferous period. They were accompanied by volcanic action, and were most intense along certain lines, and especially near the junction of the Carboniferous with the older formations.

I have noticed an apparent case of unconformability between members of the Carboniferous system near Antigonish[*]. In the county of Pictou the arrangement of the beds suggests a possible unconformability of the Upper Coal-formation and the Coal-measures[†]. In New Brunswick Prof. Bailey[‡] has observed indications of local unconformability of the Coal-formation with the Lower Carboniferous. But the strict conformability of all the members of the Carboniferous series in the great majority of cases, shows that these instances of unconformability are exceptional. In the section at the Joggins, more especially, the whole series presents a regular dip, diminishing gradually from the margin to the middle line of the trough, where the beds become horizontal.

The most gradual and uniform oscillations of level must, however, be accompanied with irregularities of deposition and local denudation; and phenomena of this kind are abundantly manifest in the Carboniferous strata of Nova Scotia. I have described in 'Acadian Geology' a bed in the Pictou coal-field which seems to be an ancient shingle-beach, extending across a bay or indentation in the coast-line of the Carboniferous period[§]. At the Joggins many instances occur of the sudden running out and cutting off of beds[||], and Mr. Brown has figured a number of instances of this kind in the Coal-formation of Sydney[¶]. They are of such a character as to indicate the cutting action of tidal or fluviatile currents on the muddy or sandy bottom of shallow water. In some instances the layers of sand and drift-plants filling such cuts suggest the idea of tidal channels in an estuary filled with matter carried down by river-inundations. Even the beds of coal are by no means uniform when traced for considerable distances. The beds which have been mined at Pictou and the Joggins show material differences in quality and associations; and small beds may be observed to change in a remarkable manner, in

* Quart. Journ. Geol. Soc. vol. i. p. 32.
† Ibid. vol. x. p. 42; Acadian Geology, p. 249.
‡ Report on Geology of Southern New Brunswick. p. 118.
§ Quart. Journ. Geol. Soc. vol. x. p. 45. || Ibid. vol. x. p. 12.
¶ Ibid. vol. vi. p. 125 et seq.

their thickness and in the materials associated with them, in tracing them a few hundreds of feet from the top of the cliff to low-water mark on the beach. I have no doubt that, could we trace them over sufficiently large areas, they would all be found to give place to sandstones, or to run out into bituminous shales and limestones, according to the undulations of the surfaces on which they were deposited, just as the peaty matter in modern swamps thins out toward banks of sand, or passes into the muck or mud of inundated flats or ponds.

3. *Geological Cycles.*—The foregoing considerations bring, in a very distinct manner, before us two different, and at first sight irreconcileable, general views which we may take of any given geological period. *First*, we must regard every such period as presenting during its whole continuance the diversified conditions of land and water with their appropriate inhabitants; and *secondly*, we must consider each such period as forming a geological cycle, in which such conditions to a certain extent were successive. As we give prominence to one or the other of these views, our conclusions as to the character of geological chronology must vary in their character; and in order to arrive at a true picture of any given time, it is necessary to have both before us in their due proportion.

We know that the marine animals of the Lower Carboniferous seas continued to exist in the time of the Coal-formation, and that some of them survived until the Permian period, proving to us the existence of deep seas even in that age which we regard as specially characterized by swampy flats supporting land-plants. In like manner we know that some of the species of land-plants found in the lowest Coal-measures continued to exist in the time of the Upper Coal-formation, proving that there was some land suitable for them throughout the epoch of the deep-sea limestones. Regarded from this point of view, any exceptional beds with land-plants in the marine parts of the formation, or beds with sea-shells in the parts where land-conditions predominate, acquire a special interest; and so likewise do regions in which, as in some parts of the Appalachian Coal-field, the marine limestones are absent, and those in which, as in some parts of the Western States, marine conditions seem to have continued throughout the whole period. In Nova Scotia, so far as my present knowledge extends, the marine limestones of the Lower Carboniferous cut off the flora of the Lower Coal-measures, apparently by a long interval of time, from that of the Middle C..i.. formation; and in like manner the fossils of the marine limestones cease at the time of the Millstone-grit, and only in one instance, that of a small bed of limestone near Wallace Harbour, partially reappear in the Upper Coal-formation[*]. I have, however, ascertained that the Marine Limestones may be divided into an upper and a lower member, and that there is some reason to suppose that in some parts of Nova Scotia, where the true Coal-measures are not developed, the upper member may in part, at least, represent them[†].

[*] Acad. Geol. p. 183; Quart. Journ. Geol. Soc. vol. ii. p. 133.
[†] Quart. Journ. Geol. Soc. vol. xv. pp. 63 *et seq.* My friend Mr. C. F. Hartt, who

I

On the other hand, I have not as yet been able to bridge over the gulf which separates the flora of the Lower Carboniferous Coalmeasures from that of the Middle Coal-formation, an interval which may include much of the " Lower Coal-measures" of Rogers in the Pennsylvania Coal-field.

Turning to that broader view which takes the prevalent conditions of each portion of the period as characteristic, notwithstanding the local existence of dissimilar conditions, we not only find, as already stated, that the sequence in Nova Scotia coincides generally with that in other parts of America and in Europe, but that, viewed in this aspect, the Carboniferous period constitutes one of four great physical cycles, which make up the Palæozoic age in Eastern America—and each of which was characterized by a great subsidence and partial re-elevation, succeeded by a second and very gradual subsidence.   Viewed in this way, the Lower Carboniferous conglomerate and Lower Coal-measures correspond analogically with the Oriskany Sandstone, the Oneida and Medina Sandstones, and the Potsdam and Calciferous.   The Carboniferous Limestone corresponds with the Corniferous Limestone, the Niagara Limestone, and the Trenton group of limestones.   The Coal measures correspond with the Hamilton group, the Salina group, and the Utica Shale.   The Upper Coal-formation corresponds with the Chemung, the Lower Helderberg, and the Hudson-River groups.   The Permian is not represented in Eastern America; but as developed in Europe it clearly constitutes a similar cycle.   These parallelisms, which deserve more attention from geologists than they have yet received, may be tabulated thus *:—

Tabular View of Cycles in the Palæozoic Age in Eastern America.
(The several formations are arranged in descending order.)

Character of group.	Lower Silurian.	Upper Silurian.	Devonian.	Carboniferous.
Shallow, subsiding marine area, filling up with sediment .... Elevation, followed by slow subsidence, land-surfaces, &c.	Hudson-River group. Utica shale ..	Lo rer Helderberg group. Salina group..	Chemung gr... Hamilton gr...	Upper Coalformation. Coal-measures.
Marine conditions; formation of limestones, &c. ..	Trenton, Black R. and Chazy limestones.	Niagara and Clinton limestones.	Corniferous limestone.	Lower Carboniferous limestone.
Subsidence; disturbances; deposition of coarse sediment	Potsdam and Calciferous sandstones.	Oneida and Medina sandstones.	Oriskany sandstone.	Lower Coalmeasures and conglomerate.

In the Permian of Europe, the Stinkstein, the Rauchwacke, the Zechstein, and the Rothliegendes might form a fifth parallel column.

has more recently studied the Marine Limestones, has obtained facts which seem to indicate the possibility of a more minute subdivision than any hitherto attempted of these beds.

* Dr. Sterry Hunt has directed attention to them in a paper " On Bitumens," 'Silliman's Journal ' [2], xxxv. p. 166, and in the 'Geology of Canada,' 1863, p. 627 ; and Dana refers to them in his 'Manual of Geology.' Eaton and Hall had previously noticed these parallelisms.

Of course such parallelism might be variously expressed by reckoning a smaller or larger number of groups. Independently of these
different modes of statement, however, I believe that the basis of
such comparisons exists in nature, and that it will prove possible to
subdivide geological time into determinate natural cycles, the parts
of which are analogous to those of similar cycles. A further question
to be solved is, whether such cycles corresponded in all parts of the
world, or whether, as is more likely, the earth might be divided into
areas in which in each cycle elevation and subsidence were contemporaneous. So far as the present subject is concerned, I merely
desire to show that the Carboniferous rocks of Nova Scotia represent
a complete cycle of the earth's history, and correspond in time with
the Carboniferous of Europe, and in value with the other great
divisions of the Palæozoic age.

4. *Summary of facts relating to the mode of accumulation of Coal.*
—With regard to the more special subject of this paper, I would
rather invite attention to the details to be presented under the next
head, than make any preliminary general statements. It is, however, necessary to notice here the several views which have prevailed
as to the probable accumulation of coal by driftage or growth *in
situ,* in water or on land. I have already, in previous publications[*],
stated very fully the conclusions at which I have arrived on some
portions of this subject, and I would now sum up the more important general truths as follows:—(1) The occurrence of *Stigmaria*
under nearly every bed of coal, proves beyond question that the
material was accumulated by growth *in situ,* while the character of
the sediments intervening between the beds of coal proves with
equal certainty the abundant transport of mud and sand by water.
In other words, conditions similar to those of the swampy deltas of
great rivers are implied. (2) The true coal consists principally of
the flattened bark of Sigillarioid and other trees, intermixed with
leaves of ferns and *Cordaites,* and other herbaceous débris, and with
fragments of decayed wood constituting " mineral charcoal," all
these materials having manifestly alike grown and accumulated
where we find them. (3) The microscopical structure and chemical
composition of the beds of Cannel-coal and earthy bitumen, and of
the more highly bituminous and carbonaceous shales, show them to
have been of the nature of the fine vegetable mud which accumu-
lates in the ponds and shallow lakes of modern swamps. When
such fine vegetable sediment is mixed, as is often the case, with
shales, it becomes similar to the bituminous limestone and calcareo-
bituminous shales of the Coal-measures. (4) A few of the underclays which support beds of coal are of the nature of the vegetable
mud above referred to; but the greater part are argillo-arenaceous
in composition, with little vegetable matter, and bleached by the
drainage from them of water containing the products of vegetable
decay. They are, in short, loamy or clay soils, and must have been
sufficiently above water to admit of drainage. The absence of sul-

* "On the Structures of Coal," Quart. Journ. Geol. Soc. vol. xv. Air-breathers
of the Coal Period, Montreal, 1863, p. 18.

phurets, and the occurrence of carbonate of iron in connexion with
them, prove that, when they existed as soils, rain-water, and not sea-
water, percolated them.   (5) The coal and the fossil forests present
many evidences of subaërial conditions.   Most of the erect and
prostrate trees had become hollow shells of bark before they were
finally imbedded, and their wood had broken into cubical pieces of
mineral charcoal.   Land-snails and galley-worms (*Xylobius*) crept
into them, and they became dens or traps for reptiles.   Large quantities of mineral charcoal occur on the surfaces of all the larger beds
of coal.   None of these appearances could have been produced by
subaqueous action.   (6) Though the roots of *Sigillaria* bear some
resemblance to the rhizomes of certain aquatic plants, yet structu-
rally they are absolutely identical with the roots of Cycads, which
the stems also resemble.   Further, the *Sigillariæ* grew on the same
soils which supported Conifers, *Lepidodendra*, *Cordaites*, and Ferns,
plants which could not have grown in water.   Again, with the exception, perhaps, of some *Pinnulariæ* and *Asterophyllites*, there is a
remarkable absence from the Coal-measures of any form of properly
aquatic vegetation.   (7) The occurrence of marine or brackish-
water animals in the roofs of coal-beds, or even in the coal itself,
affords no evidence of subaqueous accumulation, since the same thing
occurs in the case of modern submarine forests.   For these and
other reasons, some of which are more fully stated in the papers
already referred to, while I admit that the areas of coal-accumulation were frequently submerged, I must maintain that the true
coal is a subaërial accumulation by vegetable growth on soils wet
and swampy, it is true, but not submerged.   I would add the further
consideration, already urged elsewhere, that, in the case of the fossil
forests associated with the coal, the conditions of submergence and
silting-up which have preserved the trees as fossils must have been
precisely those which were fatal to their existence as living plants—
a fact sufficiently evident to us in the case of modern submarine
forests, but often overlooked by the framers of theories of the accumulation of coal.

It seems strange that the occasional inequalities of the floors of
the coal-beds, the sand or gravel ridges which traverse them, the
channels cut through the coal, the occurrence of patches of sand,
and the insertion of wedges of such material splitting the beds, have
been regarded by some able geologists as evidences of the aqueous
origin of coal.   In truth, these appearances are of constant occurrence in modern swamps and marshes, more especially near their
margins, or where they are exposed to the effects of ocean-storms or
river-inundations.   The lamination of the coal has also been
adduced as a proof of aqueous deposition; but the microscope shows,
as I have elsewhere pointed out, that this is entirely different from
aqueous lamination, and depends on the superposition of successive
generations of more or less decayed trunks of trees and beds of
leaves.   The lamination in the truly aqueous cannels and carbonaceous shales is of a very different character.

It is scarcely necessary to remark that in the above summary I have had reference principally to the appearances presented by the Coal-formation of Nova Scotia, and that I have no wish to undervalue the admirable researches on this subject of Brongniart, Goeppert, Hawkshaw, Beaumont, Binney, Rogers, Lesquereux, and others, whose publications on this subject I have read with interest, and have tested in their application to the phenomena presented to me in the coal-fields of Nova Scotia. I may add that in my opinion the phenomena of the Stigmaria-underclays, to which attention was first directed by Sir W. E. Logan, furnish the key to the whole question of the origin of coal, and that the comparisons of Coal-deposits, by Sir Charles Lyell, with the "Cypress-swamps" of the Mississippi perfectly explain all the more important appearances in the Coal-formation of Nova Scotia.

§ III. DETAILS OF THE CHARACTER AND FOSSIL CONTENTS OF THE SEVERAL BEDS OF COAL, AS EXPOSED IN THE SOUTH JOGGINS SECTION.

1. *Introduction.*—Under this heading I propose to state all the facts bearing on the origin and mode of formation of the several coals, obtained either by careful study of their outcrops on the ground, or by subsequent examination, with the aid of the microscope, of specimens collected from them. I shall follow the order of the detailed section published by Sir W. E. Logan in 1845*, including the additional points observed by Sir C. Lyell and myself in 1852, and by myself in several successive visits†, but giving in minute detail only the coals and their associated roof-beds and underclays. The sandstones and shales which constitute the mechanical filling-in between the beds of coal I shall group together in the shortest possible manner, referring to the published sections above-mentioned for details. I shall, however, mention every case of the occurrence of beds holding erect trees, and of Stigmarian underclays, as well as of beds of bituminous limestone and highly carbonaceous shale. I regard the former as being truly land-surfaces, as well as the coals, and the latter as accumulations of vegetable mud or muck which imply the contemporaneous existence in their vicinity of swamps and forests.

2. *Logan's Section* (order descending). a. *Division 1.*—This extends along the coast from Shoulie River to the vicinity of Ragged Reef, being nearly horizontal at the former place and gradually assuming a decided south-west dip towards the latter. It is 1017 feet in vertical thickness, and constitutes the upper part of the " Upper Coal-formation." It occupies the centre of the great synclinal of the western part of the Cumberland coal-area, and presents the newest beds of the Carboniferous system.

The rocks are thick-bedded white and grey sandstones, passing in some places into conglomerates with quartz pebbles, and interstratified with reddish and chocolate shales. The sandstones predominate.

Fossils are not numerous in these beds. Those found are *Dadoxy-*

* Report of Progress of Canadian Survey, 1845.
† Quart. Journ. Geol. Soc. vol. x.; also Acadian Geology, p. 128 *et seq.*

lon materiarium, of which there are many drifted trunks in the sandstones in a blackened and calcified condition, *Calamites Suckovii,
C. Cistii, Calamodendron approximatum, Lepidodendron undulatum,
Lepidophloios parvus*, and *Stigmaria ficoides*. As in the Upper Coal-
formation of Pictou, trunks of Conifers and Calamites are the most
abundant fossils.

h. *Division 2.*—This occurs at Ragged Reef and its vicinity. Its
thickness is 650 feet. It constitutes the lower part of the Upper
Coal-formation.

The rocks are white and grey sandstone with occasional reddish
beds, and red and grey shales. The sandstones and shales are nearly
in equal proportions. Underclays, or soils supporting erect plants,
probably *Sigillariæ*, occur at two levels.

Fossils are not numerous. Those collected were *Sigillaria scutellata* and *Stigmaria ficoides, Calamites Suckovii, Sphenopteris
hymenophylloides, Alethopteris lonchitica, Cyclopteris heterophylla* (?),
Beinertia Gæpperti, and portions of the strobiles of two species of
Lepidophloios, namely *Lepidophyllum lanceolatum* and *L. trinerve*.

c. *Division 3.*—This extends in descending order from the vicinity
of Ragged Reef to M'Cairn's Brook. Its thickness is 2134 feet. It
includes the upper part of the "Middle Coal-formation," and is
perhaps equivalent, in part at least, to the Upper Coal-measures of
Great Britain, and to the Upper Coal-formation of American authors.

It includes 1009 feet of sandstone, almost all of which is grey,
and 912 feet of grey and reddish shale and clay. It contains 22
beds of coal, all of small thickness, and most of them of coarse
quality. Below I give each bed of coal in detail, with its roof and
floor and its fossils; and the intervening mechanical beds in brac-
kets. The thickness of the roofs and floors is included in that stated
for the intervening beds.

		ft. in.
(Carbonaceous shale, grey understone, with *Stigmaria* and grey shale)		7 0
Coal-group 1 ...{ Grey argillaceous shale.		
Coal 1 inch.		0 1
Grey argillaceous underclay, *Stigmaria.*		

The roof holds abundance of *Alethopteris lonchitica*. The
coal is coarse and earthy, with much epidermal and bast
tissue[*], vascular bundles of ferns, and impressions of *Sigillaria* and *Cordaites*. It is a compressed vegetable soil or
dirt-bed, resting on an argillaceous subsoil with rootlets of
Stigmaria.

(Grey and reddish sandstones and grey and red shales with ironstone nodules)		281 6
Coal-group 2 ...{ Reddish argillaceous shale.		
Coal 1 inch.		
Carbonaceous shale 4 inches... }		0 6
Coal 1 inch.		
Reddish underclay, *Stigmaria.*		

---

[*] For explanation as to the nature of these and other structures in the coal,
see under § IV., below.

The coal is coarse, earthy, and shaly. It contains *Cordaites*, fern stipes, and bast tissue.

	ft.	in.
(Reddish shale and grey sandstone, the latter seen in the cliff to thin out and give place to reddish shale)	53	9
Coal-group 3 ... { Grey sandstone. *Coal* 1 inch....	0	1
Grey and reddish sandy understone, *Stigmaria*.		

The coal is coarse and shaly. No fossils were observed, except stumps and rootlets of *Stigmaria* in the underclay.

	ft.	in.
(Reddish grey shale and grey sandstone)	6	0
Reddish grey shale.		
Coal-group 4 ... { *Coal* 2 inches...	0	2
Grey and reddish argillaceous underclay, *Stigmaria*.		

The coal is coarse and earthy. No fossils were observed, except *Stigmaria* rootlets in the underclay. This and the last coal are to be regarded merely as fossil vegetable soils or dirt-beds.

	ft.	in.
(Grey sandstone and grey and reddish shale. One underclay, and erect *Calamites* in the lowest bed)	239	6
Grey argillaceous shale.		
Coal-group 5 ... { *Coal* 2 inches	0	2
Grey argillo-arenaceous underclay, *Stigmaria*.		

The coal is filled with leaves of *Cordaites borassifolia*, dividing it into thin papery layers. The underclay has many large branching roots of *Stigmaria*.

	ft.	in.
(Grey shale and sandstone)	19	0
Grey arenaceous shale.		
Coal-group 6 ... { *Coal* 3 inches	0	3
Grey argillo-arenaceous underclay, *Stigmaria*.		

This coal is composed of flattened bark of *Sigillaria*, of which there are many layers in the thickness of the bed. The species are not distinguishable.

	ft.	in.
(Grey sandstone and shale. One underclay with *Stigmaria*)...	12	6
Grey argillaceous shale.		
Coal 1 inch.		
Grey argillaceous underclay, *Stigmaria*, 1 ft. 6 in.		
Coal-group 7 ... { Coal 2 inches.		
Grey argillaceous underclay, *Stigmaria*, 4 inches.		
Coal 1 inch.		
Grey argillaceous underclay, *Stigmaria*	2	2

This is an alternation of thin coarse coals or fossil vegetable soils with *Stigmaria* subsoils. The roof-shale contains erect *Calamites*, which seem to have been the last vegetation which grew on the surface of the upper coal.

	ft.	in.
(Grey and reddish sandstones and shales)	73	0
Red and grey shale.		
Coal-group 8 ... { *Coal* 1 inch...	0	1
Grey hard underclay, *Stigmaria*.		

This coal contains flattened trunks of *Sigillaria scutellata*, or an allied species, and of other *Sigillariæ*, also

abundance of vascular bundles of ferns and portions of   ft. in.
epidermal tissues of different plants.

		ft.	in.
	(Grey sandstone and red and grey shales. *Stigmaria* in the upper bed, and prostrate *Sigillaria* and *Cordaites* in some of the sandstones and shales)...	490	0
Coal-group 9 ...	{ Grey argillaceous shale, ironstone nodules. { *Coal* 3 inches .... { Argillo-arenaceous underclay, *Stigmaria*.	0	3

The roof of this coal holds prostrate *Sigillariæ* of three
species and *Cordaites borassifolia*. The coal is hard and
shining, with impressions of flattened *Sigillariæ*, also of
Cordaites, Asterophyllites, Carpolites, and vascular bundles
of ferns.

		ft.	in.
	(Underclay and reddish grey shale)	6	0
Coal-group 10...	{ Reddish grey shale. { *Coal* and coaly shale 8 inches. { Grey argillaceous underclay, nodules of ironstone, and *Stigmaria* 2 feet. { *Coal*, stony and compact, 2 inches .... { Grey argillaceous underclay, *Stigmaria*.	2	10

The roof-shale has obscure impressions of plants, apparently petioles of ferns. The upper coal is thinly laminated
and full of leaves of *Cordaites* and ferns, among which is
Alethopteris lonchitica. The lower coal is compact, resembling cannel, and has many vascular bundles of ferns. It
seems to be composed of herbaceous matter macerated in
water and mixed with mud.

		ft.	in.
	(Grey sandstone and shale with nodules of ironstone)	23	0
Coal-group 11...	{ Grey argillaceous shale. { *Coal*, shaly, 3 inches.............................. { Arenaceous underclay, *Stigmaria*.	0	3

An erect ribbed *Sigillaria* appears in the roof-shale. The
coal contains many flattened *Sigillariæ*, also *Trigonocarpa*,
Cordaites, and vascular bundles of ferns.

		ft.	in.
	(Arenaceous understone with ironstone nodules and *Stigmaria*, and carbonaceous shale)..................	7	0
Coal-group 12...	{ Carbonaceous shale. { *Coal* 2 inches .... { Argillaceous underclay, ironstone, and *Stigmaria*.	0	2

This coal is hard and laminated, with many vascular
bundles of ferns upon its surfaces.

		ft.	in.
	(Grey sandstone and grey argillaceous shale)............	12	0
Coal-group 13...	{ Grey argillaceous shale. { *Coal* 7 inches .;...................................... { Grey argillaceous underclay, ironstone, and *Stigmaria*.	0	7

The roof contains erect stumps, not distinctly marked.
The coal has indications of bark of *Sigillaria*, and is hard
and shining, with a coarse earthy layer in the middle.

		ft.	in.
	(Grey shale)	7	0
Coal-group 14...	{ Grey shale, as above. { *Coal* 4 inches	0	4

		ft.	in.
Coal-group 14...	Grey argillo-arenaceous underclay, ironstone, and *Stigmaria*	1	6
	Coal 2 inches	0	2
	Grey argillaceous underclay, ironstone, and *Stigmaria*.		

The upper coal has impressions of bark of trees and *Cordaites*, especially in its upper part.

	(Grey and reddish shale and grey sandstone, with Stigmarian soils at two levels)	52	0
Coal-group 15...	Grey shale.		
	Carbonaceous *shale* 2 inches.		
	Argillaceous underclay, ironstone, and *Stigmaria*...	1	0
	Coal 1 inch	0	3
	Argillaceous underclay, ironstone, and *Stigmaria*....		

The upper shaly bed is a coal interlaminated with shale, which enables the nature of the coaly matter to be ascertained. It contains flattened *Sigillariæ* of several species, *Calamites*, *Cordaites*, *Cyperites*, leaves of *Sigillaria*, and *Lepidophylla*. The clay parting is the roof of the lower coal, and contains *Cyperites* and *Cordaites*. It has been converted into an underclay by the growth of *Sigillaria* upon it in the formation of the upper bed of coaly shale. The lower coal is compact, but showed an impression of a Calamite.

	(Grey sandstone and grey and reddish shale, ironstone nodules)	16	0
Coal-group 16...	Grey argillaceous shale.		
	Coal and carbonaceous shale 2 inches	0	2
	Reddish argillaceous underclay, ironstone, and *Stigmaria*.		

The roof supports an erect tree, a *Sigillaria* 8 feet high and 1 foot in diameter. It is also rich in *Cyperites*[*], *Cordaites*, and *Calamites*. The coal contains *Calamites* and und also discigerous tissue of Conifers or *Sigillaria*.

	(Grey sandstones and reddish and grey shales, with several Stigmarian underclays and coaly films or thin vegetable soils. One of the underclays supports large stumps of *Sigillaria*, with *Cyperites*, *Cordaites*, and *Lepidodendron* in the bed around their bases)	38	6
Coal-group 17...	Red and grey argillaceous shale.		
	Coal 1 inch.		
	Grey argillo-arenaceous underclay, *Stigmaria*, 4 ft.		
	Coal 4 inches.		
	Carbonaceous shale 4 inches.		
	Coal 1 inch	4	10
	Grey arenaceous underclay, *Stigmaria*.		

The upper layer of coal consists in part of leaves of *Cordaites*. The middle layer has much *Cordaites* and *Cyperites*.

	(Underclay and grey shale)	2	3
Coal-group 18...	Grey shale, as above.		
	Coal 3 inches	0	3
	Grey arenaceous underclay, *Stigmaria*.		

---

* By this term I continue, for convenience, to designate the leaves of *Sigillariæ*.

		ft.	in.
(Grey sandstone, and red and grey shale. Stigmarian soils at two levels)		26	0
Coal-group 19... { Reddish shale.			
Coal 1 inch.		0	1
Red argillaceous underclay, *Stigmaria*.			

The roof contains an erect *Sigillaria*. The coal and that of the previous bed were not well seen.

		ft.	in.
(Grey sandstone and red and grey shales, with many drift-trunks and erect *Sigillaria* at four levels)...		222	0
Coal-group 20... { Grey shale.			
Coal 1 inch.		0	1
Red and grey underclay, *Stigmaria*.		36	0

This coal contains much *Cordaites*.

		ft.	in.
(Grey and red shales and grey sandstone. One Stigmarian soil, and resting on it carbonaceous shale with *Cyperites*)		16	3
Coal-group 20a* { Grey shale.			
Coal 2 inches.			
Underclay, *Stigmaria*, 2 inches.			
Coal 1 inch.			
Underclay 1 inch, *Stigmaria*.			
Coal 3 inches		0	9
Argillaceous underclay, ironstone, and *Stigmaria*.			

These coals contain mineral charcoal, showing scalariform and epidermal tissues. The coals are impure, and were probably concealed at the time of Sir W. E. Logan's visit.

		ft.	in.
(Sandstone and red and grey shale, with one Stigmarian soil)		103	3
Coal-group 21... { Red shale.			
Coal and carbonaceous shale, 2 inches.		0	2
Grey argillaceous underclay, *Stigmaria*.			

In the bed above the roof-shale are erect *Calamites*. The coal is an uneven or irregular bed, and consists of flattened *Sigillariæ*, *Cyperites*, *Cordaites*, and ferns.

		ft.	in.
(Grey and reddish sandstones and shales, with drift-trunks of *Dadoxylon materiarium*, *Sigillaria*, and *Calamites*)		334	0
Coal-group 22... { Grey and red shale, nodules of ironstone.			
Coal and carbonaceous shale, 2 inches		0	2
Grey argillo-arenaceous underclay, *Stigmaria*.			

This coal consists of flattened bark of *Sigillaria* with *Cordaites*, and vascular bundles of ferns. It contains also remains of fishes. Among these was found a tooth of *Ctenoptychius*. The underclay includes stumps of *Stigmaria*, as well as rootlets.

		ft.	in.
(Grey sandstone and shale with one Stigmarian soil supporting erect stumps of *Sigillaria*)		68	0

Total thickness of Division 3, according to Logan's measurements 2159 8

d. *Division 4.*—This division of the section extends from M'Cairn's Cove to the end of the high cliff beyond "Coal-mine Point." It

---

* I designate in this way coal-groups not noticed in Logan's section.

corresponds to the lower part of the Middle Coal-formation, and probably to the Lower Coal-formation of some American authors. Its thickness, according to the measurements of Sir William E. Logan, is 2539 feet. It is remarkable for the prevalence of grey sandstones and grey and dark-coloured shales. It constitutes the part of the section re-examined by Sir C. Lyell and myself in 1852; and in the memoir which I subsequently published it is divided into 27 groups or subdivisions. For facility of reference those groups are indicated by the Roman numerals in the following pages, beginning with the highest group, XXVII.

XXVII.

	ft. in.
Bituminous limestone and calcareo-bituminous shale 4 feet.	
Coal-group 1 ... { Coal 1 foot .... :	5 0
Grey argillo-arenaceous underclay, *Stigmaria*.	

The roof has *Naiadites carbonarius* and *N. e. gatus*, *Spirorbis carbonarius*, scales of *Rhizodus*, and obscure vegetable fragments. The coal contains flattened *Sigillariæ*, *Cordaites*, *Alethopteris lonchitica*, *Cyperites*, *Calamites Nova-scotica*, and many vascular bundles of ferns.

	ft. in.
(Grey sandstone and shale with six underclays and erect *Sigillaria* at two levels; also a thin shale with *Naiadites*, *Cythere*, *Calamites*, and *Cordaites*. One of the sandstones has scales and tooth of a large fish (? *Rhizodus*) and plants covered with *Spirorbis*)	50 0
Grey argillaceous shale.	
Coal 1 inch.	
Clay 3 inches.	
Coal 1 inch.	
Coal-group 2 ... { Clay 1 inch.	
Coal 1 inch.	
Shale 4 inches.	
Coal 3 inches	1 2
Grey argillo-arenaceous underclay, *Stigmaria*.	

The roof has numerous vegetable fragments and flattened *Sigillariæ* and *Calamites*. One of the coals contains mineral charcoal, showing bast tissue, scalariform tissue, and fragments of epidermis. The lower coal has bark of *Sigillaria*, *Stigmaria*, and *Cyperites*, also numerous *Trigonocarpa* and vascular bundles of ferns. The clay partings and the underclay have obscure rootlets, probably of *Stigmaria*.

	ft. in.
(Arenaceous underclay and shale with remains of *Stigmaria*.)	
Grey argillaceous shale.	
Coal-group 3 ... { Coal 3 inches	0 3
Hard argillo-arenaceous underclay, *Stigmaria*.	

The roof has stumps of *Sigillariæ*, erect and with roots of *Stigmaria* descending among them from the bed above. The coal, which is coarse and earthy, has vascular bundles of ferns, scalariform vessels, bast tissue, and scales and

spines of fishes (*Palæoniscus*, &c.), with coprolitic matter.   ft. in.
The underclay shows abundant Stigmarian rootlets.

(Underclay and grey arenaceous shale).....................   6 0

Coal-group 4 ...
> Grey argillaceous shale.
> Coal 9 inches.
> Carbonaceous shale 6 inches.
> Coal 1 inch.
> Carbonaceous shale 4 inches.
> Coal 1 inch.
> Carbonaceous shale 8 inches.
> Coal 2 inches.
> Grey shale 1 foot 7 inches.
> Coal 8 inches 4 10
> Argillo-arenaceous underclay, *Stigmaria*.

The roof contains obscure flattened plants. The coal is hard or shaly, with vascular bundles of ferns and bast tissue. The carbonaceous shales yield *Cordaites borassifolia, Alethopteris lonchitica, Calamites, Sigillaria*, and *Cyperites*. The grey shale parting has erect stumps, apparently of *Sigillaria*. The upper shales and coals are very pyritous, and decompose when exposed to the weather—an indication that sea-water had access to these beds while the vegetable matter was still recent.

XXVI.

(Grey argillaceous sandstone and red and grey shale, with two Stigmarian soils. Footprints, probably of *Dendrerpeton*, and rain-marks occur in these beds; and it was in one of them that Mr. Marsh discovered the vertebræ of *Eosaurus Acadianus*) .   82 0

XXV.

Coal-group 5 ...
> Bituminous limestone 2 feet.
> Coal ½ inch.
> Argillo-arenaceous clay, *Stigmaria*, 6 inches.
> Coaly shale ½ inch.
> Grey argillo-arenaceous shale, ironstone nodules, *Stigmaria*, 1 foot 6 inches.
> Coaly shale 1 inch.
> Grey shale, ironstone nodules, *Stigmaria*, 2 ft. 6 in.
> Coal 6 inches 7 2
> Argillo-arenaceous underclay, *Stigmaria*.

The bituminous limestone of the roof contains *Naiadites carbonarius* and *N. elongatus*, fish-scales, and Cyprids. The upper layer of coal contains impressions of *Sigillaria* and *Lepidodendron*, on some of which are shells of *Spirorbis*. It has epidermal tissues, vascular bundles of ferns, and reticulated vessels. The coaly shales are of the nature of coarse coals, but with numerous thin layers of shaly matter. The lower coal contains petioles of ferns and *Cordaites* matted together, and numerous *Cardiocarpa*. The two thick clay partings and the underclay are Stigmarian soils.

XXIV.

(Grey sandstone and chocolate and grey shales, with two Stigmarian soils) 147 0

XXIII.

		ft.	in.
Coal-group 6 ...	Carbonaceous shale, passing downward into bituminous limestone, 1 foot 10 inches.		
	Coal 4 inches	2	2
	Argillo-arenaceous underclay, *Stigmaria*.		

The roof contains *Naiadites carbonarius, Cythere, Spirorbis*, fish-scales, and coprolites. The coal is hard and laminated, and has on its surfaces leaves of *Cordaites* and vascular bundles of ferns. It is remarkable for containing scattered remains of a number of species of fishes, belonging to the genera *Ctenoptychius, Diplodus, Palæoniscus*, and *Rhizodus*. The underclay has rootlets of *Stigmaria*, and the bed below this has large roots of the same.

		ft.	in.
Coal-group 7 ...	(Grey sandstone and shale, the latter with nodules of ironstone. Erect trees at one level)	30	0
	Grey sandstone.		
	Coal 10 inches.		
	Carbonaceous shale 2 inches		
	Coal 10 inches.		
	Carbonaceous shale 2 inches.		
	Coal and coaly shale 2 feet 6 inches.....................	4	6
	Grey argillo-arenaceous underclay, *Stigmaria*.		

This is the bed worked at the Joggins as the "Main Seam ;" and I believe that it improves somewhat in mining it inward from the shore. The roof has afforded *Sigillaria catenoides* and other species, *Alethopteris lonchitica, Cordaites borassifolia, Lepidodendron elegans, Trigonocarpa, Naiadites, Spirorbis, Cythere*, fragments of insects (?). The mineral charcoal contains bast tissue, scalariform, epidermal, and cellular tissues. In the compact part of the coal there is dense cellular and epidermal tissue. The roof is especially rich in *Cordaites*, sometimes with *Spirorbis* adherent.

		ft.	in.
Coal-group 8 ...	(Grey sandstone and shale, with many ironstone nodules in the shale, and erect *Sigillaria* and underclays at five levels. One of the latter has large stumps of *Stigmaria* and a thin coaly layer resting on it)	68	0
	Grey shale with nodules of ironstone.		
	Coal 2 inches.		
	Grey shale 4 inches.		
	Coal 3 inches.		
	Carbonaceous shale 1 foot 3 inches.		
	Coal 1 inch.		
	Argillaceous shale, ironstone nodules, 4 feet.		
	Coal 1 foot..	7	1
	Grey argillo-arenaceous underclay, ironstone nodules, and *Stigmaria*.		

The roofs of the first and second beds in this group are among the richest in fossils in the Joggins section. They have afforded *Pecopteris lonchitica, Cyclopteris, Cyperites, Cordaites borassifolia, Cardiocarpum fluitans, Sigillaria elegans, Lepidophloios Acadianus, Lepidodendron undulatum,*

Pinnularia, Trigonocarpa, &c.; also *Diplostylus Dawsoni**,   ft. in.
Eurypterus, Cythere, Naiadites, and *Spirorbis* attached to
plants. The lower coal, called locally the "Queen's Vein,"
has in its mineral charcoal bast cells, uniporous, rari-
porous, and multiporous wood-cells, scalariform vessels,
epidermal tissue, and vascular bundles of ferns, also stipes
of ferns and bark of *Sigillaria.* The mineral charcoal occurs
principally in a thick layer near the bottom of the bed. Its
roof has trunks of *Lepidophloios, Lepidodendron,* and *Sigillaria,* fossilised by carbonate of iron. The upper part of
the lowest underclay is dark and carbonaceous, with Stigmarian rootlets.

XXII.

(Grey sandstones, grey and chocolate shales with
ironstone nodules; three underclays and erect
Calamites and *Sigillaria* in three beds) 110   0

XXI.

Coal-group 9 ... { Grey shale and ironstone nodules.
Coal and coaly shale 1 foot 3 inches   1   3
Argillaceous underclay, *Stigmaria.*

The roof contains erect *Sigillariæ, Stigmaria, Calamites,*
and *Cordaites.* The coaly shale has fern-stipes and *Cordaites.* The coal itself is coarse and shaly, and has a layer
of mineral charcoal containing bast and epidermal tissue.
There are also in the coal remains of *Calamites* and *Cordaites,* and fragments, possibly, of insects.

(Grey and reddish shales with nodules of clay-iron-
stone, and grey and reddish sandstone. One underclay supporting a coaly film, and erect trees at
two levels)   28   6

Coal-group 10... { Chocolate shale.
Coal and coaly shale 2 inches.
Coaly shale 0 inches.
Coal 4 inches   1   0
Argillo-arenaceous underclay, *Stigmaria.*

The upper coal contains flattened *Sigillariæ* and *Stigmaria.* The lower bed is hard and unequal, with curved
laminæ and obscure traces of petioles of ferns. The mineral
charcoal has bast and scalariform tissues.

XX.

(Red and grey shales and grey sandstones. Erect *Calamites* in one bed. Four underclays)   78   6

XIX.

Coal-group 11... { Chocolate shale.
Coal and coaly shale 8 inches   0   8
Argillaceous underclay, *Stigmaria.*

The roof has *Cordaites, Calamites,* and rootlets. The coal
contains much mineral charcoal with the structure of dense

* Salter, Quart. Journ. Geol. Soc. vol. xix. p. 77.

aporous bast tissue; it also contains *Cyperites* and many vascular bundles of ferns.

	(Grey sandstones and argillaceous shale. Erect trees at two levels)	37	0
Coal-group 12...	Grey shale. Coal and coaly shale 1 foot Argillaceous underclay, ironstone, and *Stigmaria*.	1	0

The roof contains erect *Sigillaria* and *Calamites*, also *Cordaites* with *Spirorbis* attached, and *Lepidodendron*. The coal has in one layer much *Cordaites*, in others it includes an immense number of specimens of *Sporangites papillata*; it has also bast tissue, epidermal tissue, and discigerous tissue.

	(Shale and sandstone, penetrated by Stigmarian rootlets, and containing in one of the shales *Lepidodendron*, *Sigillaria*, and *Carpolithes*)	13	0
Coal-group 13...	Grey shale. Coal and coaly shale Argillaceous underclay, *Stigmaria*.	0	6

The roof has much *Cordaites*. The shaly portions of the coal contain *Sigillaria elegans*, *Alethopteris lonchitica*, *Cordaites borassifolia*, *Lepidodendron*, *Diplotegium*, *Trigonocarpum*, *Stigmaria*, and *Sporangites glabra*, also vascular bundles of ferns and bast tissue.

XVIII.

	(Grey and red shales and grey sandstone; one of the latter with erect *Calamites* and *Sigillaria*. One underclay)	69	4

XVII.

Coal-group 13a.	Grey shale. Coal 8 inches Argillaceous underclay, *Stigmaria*.	0	8

The roof has *Cordaites* and many decayed stipes. The coal has *Cordaites* and vegetable fragments.

XVI.

	(A very thick sandstone with shales. Erect *Calamites*, footprints of reptiles, and rain-marks)	57	0

XV.

Coal-group 14...	Grey shales with ironstone. Coal 3 inches. Coaly shale 2 inches. Coal 3 inches. Underclay, *Stigmaria*, 6 feet. Coaly shale 4 inches. Underclay, *Stigmaria*, 1 foot. Coaly shale 8 inches. Coal 2 inches Argillo-arenaceous underclay, *Stigmaria*, and ironstone.	8	10

On the roof of the upper coal is a fine-ribbed *Sigillaria* with Stigmarian roots. In the roof and shaly partings are *Sigillaria Brownii*, *S. Schlotheimiana*, and other species, Stigmaria, Lepidodendron, Calamites, Cordaites, Sporangites glabra, Alethopteris lonchitica, Sphenopteris latifolia, Pinnularia, and Cyperites; also Cythere, Naiadites, and fragments of reptilian (?) bones. The coal is pyritous, and exhibits impressions of the bark of *Sigillaria*; it contains also bast tissue, scalariform tissue of *Sigillaria*, and multiporous tissue of *Sigillaria* and *Calamodendron*.

	ft. in.
(Sandstone and shale, erect *Calamites* and *Sigillaria* with *Stigmaria*. The erect trees contain reptilian remains of the genera *Dendrerpeton*, *Hylonomus*, and *Hylerpeton*; also *Pupa vetusta*, *Xylobius Sigillariæ*, and remains of insects)	10 0
Coal-group 15... { Coaly shale. Coal 6 inches	0 6
Arenaceous underclay, *Stigmaria*.	

The erect trees above mentioned are rooted in the roof of this coal. It contains *Cyperites*, *Lepidophylla*, *Trigonocarpa* of 2 species, *Sphenophyllum*, *Alethopteris lonchitica*, *Cordaites*, and *Asterophyllites*. There are shells of *Spirorbis* on some of the plants. The coal contains layers of bark of *Sigillaria* and leaves of *Cordaites*, and much bast tissue, with scalariform, uniporous, and reticulated tissues, probably of *Sigillaria*.

(Sandstones and shales; erect *Calamites* and *Stigmaria*).....	21 0
Coal-group 15a. { Grey shale. Coal 4 inches	0 4
Argillaceous underclay, *Stigmaria*.	

The roof contains *Calamites*, *Sigillaria*, *Alethopteris lonchitica*, *Pinnularia*, *Lepidodendron*, *Cyperites*, *Sporangites*, and *Spirorbis*. One *Sigillaria* extends 30 feet without branching. The roof supports an erect tree. The coal is filled with flattened stems of *Sigillaria* lying in different directions, also flattened *Lepidodendra*; and in its mineral charcoal it has beautiful porous and scalariform tissues.

XIV.

(Grey sandstone and grey and red shales. Many prostrate trunks of *Sigillaria* and *Lepidodendron*, one underclay, and erect trees at one level)	68 0
Coal-group 16... { Shale with the aspect of underclay. Coal and coaly shale 6 inches	0 6
Argillo-arenaceous underclay, ironstone, and *Stigmaria*.	

This bed was not well exposed, and afforded no fossils.

(Grey sandstone and shale with one underclay)	25 6
Coal-group 17... { Grey shale. Coal and coaly shale 3 inches	0 3
Argillo-arenaceous underclay, *Stigmaria*.	

The roof has vegetable fragments and *Cordaites*. The

coal is hard and coarse, and contains flattened broad-ribbed
Sigillaria, Cordaites, and vascular bundles of ferns.

	ft. in.
(Shale and sandstone, erect trees at one level)	31 3

XIII.

		ft. in.
Coal-group 18...	{ Grey shale. Coal 8 inches	0 8
	Argillo-arenaceous underclay.	

The roof has an erect *Sigillaria.* The coal is shaly and
laminated. It contains much *Cordaites,* also *Lepidodendron,
Calamites,* and *Alethopteris lonchitica.* In one layer there
are *Naiadites, Spirorbis,* and scales of fishes.

		ft. in.
	(Grey sandstone and shale in several beds, *Stig- maria*)..	29 0
Coal-group 19...	{ Argillaceous shale. Coaly shale 4 feet. Bituminous limestone 2 feet 6 inches. Coal 1 inch...	6 7

The roof has *Naiadites,* scales and teeth of fishes, *Cythere,*
and *Spirorbis.* The coal is hard and coarse, with vascular
bundles of ferns and prostrate *Sigillariæ.*

		ft. in.
	(Shale and sandstone)	20 6
Coal-group 20...	{ Coaly shale 1 foot. Bituminous limestone 1 foot 6 inches. *Coal* and clay partings 2 feet 4 inches................	4 10

The roof has *Naiadites, Spirorbis* attached to plants, and
small rhombic fish-scales. The coal alternates with limestone at the top, and contains remains of *Sigillaria, Sporangites,* and vascular bundles of ferns.

		ft. in.
	(Sandstone, and grey and black shale with coaly layers)...	21 0
Coal-group 21...	{ Grey shale and calcareo-bituminous shale. *Coal* 10 inches	0 10
	Argillaceous underclay, *Stigmaria.*	

The roof has obscure vegetable fragments and *Naiadites.*
The coal contains vascular bundles of ferns, bast tissue,
uniporous cells, and scalariform and reticulated vessels.

		ft. in.
	(Grey sandstone and shale. Two underclays).........	20 0
Coal-group 22...	{ Grey shale. Coal and coaly shale 2 inches	0 2
	Argillaceous underclay, *Stigmaria.*	

This bed was not well exposed.

		ft. in.
	(Sandstone and shale, with one erect tree and two underclays)...	12 0
Coal-group 23...	{ Coaly and grey shale. *Coal* and coaly shale 4 inches. Bituminous limestone 4 inches. *Coal* and coaly shale 7 inches	1 3
	Argillo-arenaceous underclay, *Stigmaria.*	

The roof has an erect tree, also *Cordaites* and *Spirorbis.*
The shale and bituminous limestone contain *Sigillaria* and

Lepidophloios, also many large-furrowed trunks, probably old *Sigillariæ* or *Lepidodendra*.

ft. in.

XII.

		ft.	in.
	(Sandstone, shale, and calcareo-bituminous shale with three underclays)	26	0
Coal-group 24...	Calcareo-bituminous shale.		
	Coal and coaly shale 1 inch	0	1
	Argillo-arenaceous underclay, *Stigmaria*.		

This bed was not exposed.

	(Underclay and shale)	5	0
Coal-group 25...	Grey shale.		
	Coal and coaly shale 8 inches	0	8
	Argillo-arenaceous underclay, *Stigmaria*.		

The roof has *Alethopteris lonchitica, Cordaites,* and petioles of ferns. The coal shows bast tissue and remains of *Sigillaria* and *Calamites.*

	(Grey sandstone and shale, with erect *Sigillariæ* at four or five levels, and two Stigmarian underclays)	117	8
Coal-group 26...	Grey shale.		
	Coaly shale 4 inches	0	4
	Argillo-arenaceous underclay, *Stigmaria*.		

This bed was not exposed.

	(Shale and sandstone, with *Stigmaria*)	13	0
Coal-group 27...	Grey shale.		
	Coal 3 inches	0	3
	Argillo-arenaceous underclay, *Stigmaria*.		

This bed was not well exposed.

(Grey sandstone and shale, with bituminous shale and limestone, and erect *Calamites*)	64	0

XI.

		ft.	in.
Coal-group 28...	Calcareo-bituminous shale.		
	Coal and coaly shale 7 feet.		
	Underclay, *Stigmaria*, 4 feet.		
	Coaly shale 1 foot.		
	Coal 6 inches	12	6
	Arenaceous underclay, *Stigmaria*.		

This group is a series of thin coaly layers and underclays. The roof has *Naiadites carbonarius* and *N. elongatus,* also *Cythere* and scales of fishes. The coal contains bast tissue and different kinds of scalariform and epidermal tissues. In the lower bed is a coaly stump and an irregular layer of mineral charcoal, arising apparently from the decay of similar stumps.

	(Grey and carbonaceous shale and grey sandstone)	29	0
Coal-group 29...	Underclay, *Stigmaria*.		
	Coal and coaly shale 5 feet.		
	Underclay 6 feet.		
	Coal, coaly shale, and ironstone, 6 feet.		
	Coal 4 feet	21	0
	Argillaceous underclay, *Stigmaria*.		

This is a group of unusually thick beds, indicating long quiescence. The roof includes laminæ of coal, some of them composed of the bark of *Sigillaria catenoides*, also an erect *Sigillaria* rooted in the coal below. The coal and coaly shale exhibit remains of *Sigillaria, Cordaites, Lepidophyllum,* and *Cyperites*; and one layer has many hard pyritized fragments of wood. The mineral charcoal has vascular bundles of ferns, coarse scalariform tissue, and porous tissue. The underclay rests on a bed with *Naiadites.*

		ft.	in.
	(Underclay, *Stigmaria,* and grey and carbonaceous shales)	18	0
Coal-group 29a .	Shale and coaly layers. *Coal* 4 feet. Argillaceous underclay, *Stigmaria.*	4	0

The roof has obscure fragments of plants and stumps in the state of mineral charcoal. The coal shows impressions of flattened trunks, probably *Sigillariæ*. This coal contains a great variety of tissues, especially bast and scalariform of different kinds, and epidermal. My measurements in this part of the section differ somewhat from those of Sir W. E. Logan, who, I suppose, had not a good opportunity of examining the two last coals. The coal 29a is now mined by an adit from the shore, called the "new mine."

		ft.	in.
	(Sandstone and shale. One sandstone has many large erect *Sigillariæ*, some of them with rough and furrowed bark)	35	0
Coal-group 30...	Argillaceous shale and ironstone. *Coal* 4 inches. Underclay, dark-coloured, 2 feet. Coal and coaly shale 2 inches. *Coal* 3 inches. Coaly shale 2 inches. *Coal* 1 inch Soft argillaceous underclay, *Stigmaria.*	3	0

The roof has bark of *Sigillaria* preserved in ironstone. The coal is pyritous, and consists of layers of mineral charcoal alternating with bright coal; it has obscure impressions of plants and bast tissue in the mineral charcoal.

x.

		ft.	in.
	(Grey shale and sandstone. One underclay, and erect *Calamites* and *Sigillaria* at two levels)	19	0
Coal-group 31...	Grey sandstone. Coal and coaly shale 1 foot. Underclay, *Stigmaria,* 1 foot. Coaly shale 6 inches. *Coal* 2 inches Argillaceous underclay, *Stigmaria.*	2	8

The roof contains *Sigillariæ*, and the coal has flattened impressions of the same. This coal is remarkable as having a roof of sandstone. Its underclay is also peculiar. It is about 9 feet in thickness, and contains *Stigmaria* and nodules of

ironstone throughout. It rests on a bituminous limestone ft. in.
containing *Naiadites* and scales of fishes, and also large roots
of *Stigmaria* evidently *in situ*. This bed gives more colour
to the idea of *Stigmaria* having grown under water than any
other bed at the Joggins. I believe, however, that it merely
implies the drying-up of a pond or creek into a swamp, subsequently inundated at intervals with muddy water.

(Underclay and bituminous limestone, succeeded by
    sandstone and shale)......   27   8
Coal-group 32... { Grey shale.
     { *Coal* and coaly shale 2 feet 4 inches   2   4
     { Argillo-arenaceous underclay, *Stigmaria.*

This is a series of thin coaly bands alternating with shales.
The roof contains trunks of *Sigillaria, Cordaites, Alethopteris*,
and *Cyperites*. The coal has numerous flattened trunks of
Sigillaria.

(Grey and reddish sandstone and shale. Five under-
    clays, one with a film of coal and erect *Sigillariæ*
    at two levels) 149 0
Coal-group 33... { Coaly shale.
     { *Coal* 1 inch..   0   1
     { Argillaceous underclay, *Stigmaria*.

The roof has flattened trunks and vegetable fragments.
The coal is a mere soil, with remains of *Sigillaria* and *Cordaites*, and vascular bundles of ferns.

(Red and grey sandstone and shale)....................   45   0
Coal-group 33a . { Grey shale.
     { *Coal* and coaly and grey shale (underclay).
     { Argillaceous underclay, *Stigmaria*.

These layers, though not of sufficient importance to be
measured as coal-bands, are most interesting as furnishing
examples of what may be termed rudimentary coal-beds.
Each layer is plainly composed of prostrate trunks of *Sigillaria* resting on Stigmarian underclay, and mixed with *Cordaites, Alethopteris lonchitica*, and vascular bundles of ferns.
In one layer is a stump in the state of mineral charcoal.
In another there are coprolites, scales of fishes, *Spirorbis*,
and fragments of Crustaceans. In a reddish shale above
these beds there is a patch of grey sandstone interlaced with
Stigmarian roots, as if the sand had been prevented from
drifting away by a tree or stump.

(Reddish and grey sandstones and shales, with three
    or more underclays, having their coaly layers
    holding *Sigillaria*. Erect *Sigillariæ* at two levels)   240 0
Coal-group 34... { Underclay, with ironstone and *Stigmaria*.
     { *Coal* and coaly shale 2 inches.
     { Underclay, with ironstone and *Stigmaria*, 4 feet.
     { *Coal* and coaly shale 2 inches   4   4
     { Argillo-arenaceous underclay, *Stigmaria*.

Only obscure vegetable fragments were observed.

(Grey and reddish sandstone and shale, with *Stig-
    maria*)..   13 10

		ft.	in.
Coal-group 35...	Underclay with *Stigmaria*. Coaly shale 3 inches .... Red and greenish underclay, a few rootlets.	0	3

The coaly shale contains many leaves of *Cordaites borassifolia*.

(Red, grey, and dark shale, sandstone, and bituminous limestone. Three underclays and erect trees at one level)	67	9

IX.

Coal-group 36...	Bituminous limestone. Coaly shale and *coal* 3 inches. Reddish shale and ironstone, 2 feet 6 inches. *Coal* 3 inches .... Argillaceous underclay, *Stigmaria*.	3	0

The roof has *Stigmaria* in situ, and has been a soil or underclay. It also contains *Cythere*, fish-scales, coprolites, and *Spirorbis*. In the upper coaly shale are prostrate carbonized trunks.

(Reddish and grey shale, sandstone, and bituminous limestone)	21	6
Coal-group 37...	Bituminous limestone and shale. *Coal* 4 inches. Underclay 1 foot 6 inches. *Coal* 6 inches. Underclay 1 foot. Bituminous limestone 3 inches. Shale 3 inches. *Coal* 1 inch .... Underclay with *Stigmaria*.	3	11

The roof has *Stigmaria*, also fish-scales, *Naiadites*, and *Cythere*. The shales are pyritized. The coal shows only obscure fragments of plants; but *Sigillariæ* in the state of ironstone occur in some of the clays.

VIII.

(Red and grey sandstone and shale. Two underclays. Many shells of *Pupa retusta* occur in one of these, about 12 feet below the last coal).........	83	0

VII.

Coal-group 38...	Calcareo-bituminous shale. *Coal* 1 inch. Bituminous limestone 6 inches. *Coal* 2 inches. Underclay passing into chocolate shale, *Stigmaria*.		

The bituminous limestone and shale contain *Cythere*, *Naiadites elongatus* and *N. carbonarius*, coprolites, *Spiroubis*, and *Stigmaria*. The lower coal has *Sigillaria elegans*, *S. scutellata* (?), *S. Brownii*, *Alethopteris lonchitica*, *Cordaites borassifolia*, and vascular bundles of ferns.

(Red and grey shales and sandstones, and one grey limestone with *Cythere*. One underclay. Many drift trunks, among which are *Sigillaria* and *Lepidoph'oios*)	123	6

V.

		ft.	in.
Coal-group 39...	Red and grey shale, with ironstone.		
	Coal ½ inch	0	0½
	Grey underclay, with *Stigmaria*, resting on bituminous limestone, with *Stigmaria* and *Cythere*.		

This thin coal consists of a layer of flattened trunks, probably of *Sigillaria*, with a quantity of mineral charcoal.

IV.

(Red and grey shales. One bed with erect *Calamites*, another with erect *Sigillaria*)	65	4

III.

Coal-group 40...	Grey shale and ironstone. Bituminous limestone and shale, with coaly films, 7 inches. Underclay 1 foot. *Coal* 1 inch. Coaly shale 3 inches. Underclay 1 foot. Bituminous limestone 6 inches. *Coal* and coaly shale 2 inches	3	7
	Argillaceous underclay, ironstone, and *Stigmaria*.		

The bituminous limestone and shale have *Naiadites, Cythere, Spirorbis*, scales of fishes, and coprolites, and a large spine of *Gyracanthus*, also roots of *Stigmaria*. The upper underclay holds carbonized erect trunks. The lower coal has vascular bundles of ferns and *Cordaites*. The roof supports erect stumps.

Coal-group 41...	(Underclay with ironstone nodules)	5	0
	Underclay as above.		
	Calcareo-bituminous shale and films of coal	3	4
	Argillaceous underclay, *Stigmaria*.		

The bituminous limestone has *Naiadites carbonarius, Cythere*, coprolites, and *Spirorbis*. The roof has prostrate *Sigillariæ* converted into coaly layers. The underclay has distinct stumps of *Stigmaria*.

Coal-group 42...	(Shales with *Stigmaria* and ironstone, sandstones, bituminous limestone, and carbonaceous shale at bottom)	14	4
	Bituminous limestone.		
	Coal 3 inches.		
	Shale 1 foot.		
	Coal 1 foot.		
	Underclay, *Stigmaria*, 1 foot.		
	Coal 2 inches	3	5
	Dark argillaceous underclay, *Stigmaria*.		

The roof contains *Naiadites, Cythere*, and coprolites. The coal is coarse, pyritous, and shaly, and has bark of *Sigillaria, Calamites*, and vascular bundles of ferns. It seems to be the edge of a bed, as it thins rapidly in the direction of the bank, or to the east.

II.   ft. in.

	(Reddish shale and sandstone with one underclay	35	0
Coal-group 43...	⎰ Reddish underclay with *Stigmaria*. ⎨ Coaly shale 1 inch .... ⎱ Reddish underclay, *Stigmaria*.	0	1

This bed diminishes to a mere film towards the bank.

I.

		ft	in
	(Reddish, grey, and dark shales and sandstone, *Stigmaria* in some beds, and erect *Sigillariæ* and *Calamites* at one level)	63	
Coal-group 44...	⎧ Grey shale with ironstone. ⎪ Bituminous limestone and shale with ironstone, 10 ft. 1 in. ⎪ *Coal* ¼ inch. ⎪ Bituminous limestone, *Stigmaria*, ½ inch. ⎨ *Coal* 5 inches. ⎪ Bituminous limestone, *Stigmaria*, 2 inches. ⎪ *Coal* 1 inch. ⎪ Bituminous limestone, *Stigmaria*, 2 inches. ⎪ Coal ¼ inch .... ⎩ Argillo-arenaceous underclay, traces of rootlets.	11	0½

The bituminous limestone has scales of fishes, *Spirorbis*, and *Cythere*. The coal has *Cordaites* and vascular bundles of ferns.

	(Red and grey sandstone and shale. One underclay, and erect *Calamites* at one level)	98	6
Coal-group 45...	⎧ Reddish shale. ⎪ Carbonaceous shale, 10 inches. ⎪ Coaly matter ¼ inch. ⎨ Hard underclay, *Stigmaria*, 2 feet. ⎪ Coaly matter ¼ inch. ⎪ Underclay, *Stigmaria*, 7 feet. ⎪ Coal 3 inches .... ⎩ Arenaceous underclay, *Stigmaria*.	10	2

In the roof of the lower coal is an erect tree. The coal has vascular bundles of ferns, remains of fern-leaves, and bast tissue. The underclay has many coaly films, apparently flattened bark of trees.

Reddish and grey sandstone and shale 5 6

Total thickness of Division 4, according to Logan's measurements...... 2539 1

e. *Division 5.*—This consists of reddish shales and red and grey sandstones. It contains no coal, and is poor in fossils, only a few drifted trunks appearing in the section. It corresponds to the upper part of the Millstone-grit series. Its thickness, according to the measurements of Sir William E. Logan, is 2082 feet.

f. *Division 6.*—This may be regarded as the middle of the Millstone-grit series. It constitutes a sort of false coal-formation, separated from the Middle Coal-formation by the barren beds of Division 5. It contains nine small or rudimentary coal-beds, which, however, are not well seen in the section, and have afforded few facts of interest. It has many thick and coarse sandstones and much red shale, with comparatively few dark-coloured beds. Its total thickness is stated by Sir W. E. Logan at 3240 feet.

Though this group contains little coal, it is to be observed that it has many underclays, indicating soils which supported forests of *Sigillaria*, and that erect *Sigillariæ* occur very near the base of the division. The absence of important beds of coal is therefore due to the local physical conditions, and not to the want of the necessary vegetation.

		ft.	in.
(Sandstones and shales with many drifted trunks of *Dadoxylon*)		539	7
Coal-group 1 ...	Blackish grey shale. Calcareous shale 1 foot. Black shale 3 feet. *Coaly shale* 2 inches Argillo-arenaceous underclay, *Stigmaria*.	4	2
	(Red and grey sandstone and shale and concretionary limestone, trunks of *Dadoxylon* and other trees. One underclay)	160	1
Coal-group 2 ...	Grey shale. *Coaly shale* 1 inch Reddish and grey underclay; *Stigmaria*.	0	1
	(Series of underclays with *Stigmaria*. The beds are reddish or grey, and arenaceous)	19	1
Coal-group 3 ...	Reddish shale. *Coaly shale* 1 inch. Greenish shale 6 inches. *Coaly shale* 1 inch. Greenish shale 2 feet 6 inches. *Coaly shale* 3 inches. Greenish shale 1 inch. *Coal* and coaly shale 3 inches Argillo-arenaceous underclay, *Stigmaria*.	3	9

The coal contains bast tissue and reticulated, porous, and scalariform tissues of *Sigillaria* and *Calamodendron*.

		ft.	in.
	(Series of underclays as before)	12	0
Coal-group 4 ...	Underclay, *Stigmaria*. *Coal* and coaly shale 3 inches Argillo-arenaceous underclay, *Stigmaria*.	0	3
	(Series of underclays as before)	24	0
Coal-group 5 ...	Grey shale. *Coaly matter* ½ inch Greenish underclay, *Stigmaria*.	0	½
	(Underclay and sandstone, the latter with an erect *Sigillaria*)	10	0
Coal-group 6 ...	Sandstone (erect *Sigillaria* as above), *Coaly shale* 3 inches Argillo-arenaceous underclay, *Stigmaria*.	0	3
	(Fifteen feet of underclay, under which a thick sandstone with great quantities of drifted trunks of *Dadoxylon* and *Sigillaria*. Below this alternations of grey and red sandstone and shale)	210	10
Coal-group 7 ...	Grey sandstone. Bituminous limestone 3 inches. Grey shale 3 feet. Grey limestone 2 inches. *Coaly shale* 6 inches. Bituminous limestone 3 inches. *Coaly shale* 1 foot. *Coal* 1 inch Argillo-arenaceous underclay, *Stigmaria*.	5	3

The lower bituminous limestone contains *Naiadites ovalis,* ft. in.
Cythere, and scales of Lepidoid fishes. The lower coal has
much *Cyperites* and bark of *Sigillaria,* also bast tissue in
mineral charcoal.

(Thick beds of grey sandstone and grey shale, with drifted trunks of *Dadoxylon, Sigillaria,* and *Calamites,* and leaves of *Cordaites*)	532	0
Coal-group 8 ... { Grey shale. Coal ¼ inch	0	¼
Argillo-arenaceous underclay, *Stigmaria.*		

This coal is laminated, the laminæ being bark of *Sigillariæ.* The underclay is very rich in *Stigmaria.*

(Grey sandstone with grey and red shale. Many drifted trunks of *Sigillaria* and *Calamites,* and an erect *Sigillaria* in the lowest bed of sandstone)...	1224	0
Coal-group 9 ... { Grey shale. Coaly matter and carbonaceous shale	0	2
Argillo-arenaceous underclay, *Stigmaria,* and iron-stone.		
(Grey and red sandstone and shale and calcareous bands, some of them bituminous. Near the middle a thick band of laminated black shale with *Naiadites lævis, Cyperites,* and *Lepidostrobus.* Drifted *Calamites* in the sandstone)	496	4

Total thickness, according to Logan 3240 9

g. *Division 7.*—This division consists principally of red and chocolate shales with red and grey sandstone, arenaceous conglomerates,
and thin beds of concretionary limestone. It may be regarded as the
base of the Millstone-grit formation. Its thickness is stated by Sir
W. E. Logan at 650 feet.

No fossils, other than carbonized fragments of plants, have been
found in this division.

h. *Division 8.*—This division consists of reddish shales with green-
ish and red sandstone, grey shale, grey compact limestone, and gypsum. It may be regarded as the upper part of the Lower Carboniferous
formation; and almost immediately under its lowest beds there are
marine limestones with *Productus cora* and other characteristic Lower
Carboniferous fossils.

Only fragments of plants, often replaced by sulphuret of copper,
have been found in this division. Its thickness is stated by Logan
at 1658 feet.

§ IV. REMARKS ON THE ANIMALS AND PLANTS WHOSE REMAINS OCCUR
IN THE COAL.

1. *Introduction.*—Under this heading I shall, in the first place,
present a tabular view of the relative frequency of occurrence of the
several genera in the beds of coal and their roof-shales, without
reckoning the almost invariable occurrence of *Stigmaria* in the underclays, which is of course to be taken as an indication of the existence
of Sigillarioid trees in connexion with the growth of the coal.

Tho number of coals reckoned may vary according to the manner
in which the several layers are grouped; but as arranged in tho
above sectional list it amounts to eighty-one in all. Of these, 23 are
found in Division 3 of Logan's section, being the upper member of
the Middle Coal-formation; 49 are found in Division 4 of Logan's
section, being the lower member of the Middle Coal-formation; 9
occur in Division 6 of Logan's section, or in the equivalent of the
Millstone-grit. In the latter group few of the coals were sufficiently well exposed to enable a satisfactory examination to be made.
I have grouped the remains under three heads—External Forms of
Plants, Microscopic Structure of Plants, and Animal Remains—and
have arranged the forms under each in the order of their relative
frequency of occurrence.

*Table showing the Relative Frequency of Occurrence of Genera of
Plants and Animals in the Coals of the South Joggins.*

Names of Fossils.		Division 3. 23 coals.	Division 4. 49 coals.	Division 6. 9 coals.	Total. 81 coals.
Plants.					
Sigillaria..............	occurs in	13	34	2	40
Cordaites	„	15	26	...	41
Filices (mostly *Aletho- pteris lonchitica*)	„	4	17	2	23
Lepidodendron and Lepidophloios ...	„	1	15	...	16
Calamites	„	4	12	...	16
Carpolites, &c.	„	2	9	...	11
Asterophyllites	„	1	2	...	3
Calamodendron	„	...	1	...	1
Structures.					
Vascular bundles of ferns	„	8	22	...	30
Bast tissue (*Sigillaria*)	„	2	16	2	20
Epidermal tissue (*Cordaites*, &c.)...	„	6	6	...	12
Scalariform (*Sigil., Stig., Lepidod.*, &c.)	„	1	9	1	11
Discigerous (*Sigillaria* and *Dadoxylon*, &c.)	„	1	8	1	10
Reticulated (*Calamo- dendron*, Ferns, &c.)	„	...	2	1	3
Animals.					
Fishes (*Palæoniscus, Rhizodus*, &c.) ...	„	1	16	1	18
Naiadites (*Anthra- comya*, &c.).........	„	...	16	1	17
Spirorbis carbonarius	„	...	16	...	16
Cythere	„	...	13	1	14
Insects (?)	„	...	3	...	3
Reptiles (*Dendrer- peton*, &c.)	„	...	1		1
Pupa vetusta and *Xylo- bius sigillariæ*......	„	...	1	...	1

It will be observed that, whether we regard the external forms or the internal structures preserved, the predominant plants are *Sigillaria*, *Cordaites*, and Ferns, with *Lepidodendron* and *Calamites*. The substance of the coal itself, so far as its structure is preserved, may be said to be principally composed of bark of *Sigillaria* and leaves and stems of ferns and *Cordaites*. In regard to the proportions in different parts of the series, little difference exists, except that *Cordaites* and *Calamites* are rather more abundant in the upper coals, and *Lepidodendron* in the lower, while the middle of the series is the headquarters of *Sigillaria* and ferns. Remains of aquatic animals occur in connexion with a large proportion of the coals, more especially in the middle of the series. This may be explained in connexion with the theory of growth of the coal *in situ* by the following considerations:—(1) It was necessary to the preservation of the vegetable matter composing a bed of coal that it should be submerged and covered with sediment. (2) On the submergence of a swamp covered with standing trees and other vegetation, these would prevent the passage of strong currents carrying coarse detritus, and the area would be covered with fine sediment deposited in still water and under conditions favourable to certain kinds of aquatic animals. (3) When the currents carrying detritus were sufficiently powerful to uproot and sweep away the forests and the brakes of *Calamites*, they would also remove or disturb the vegetable soil. It follows that we should expect the more important coals to be covered with fine sediment containing animal as well as vegetable remains, and that beds roofed with sandstone or coarse shale must either have been of small area or sparsely covered with trees at the time of their submergence. This accounts for the otherwise anomalous circumstance that the evidences of aqueous conditions in association with the coal are proportionally more abundant in the middle than in the upper part of the Coal-measures. We may now proceed to consider the genera of plants and animals separately, in their relation to the growth of coal.

2. *Coniferous Trees.*—Four species of coniferous trees, referable to the genus *Dadoxylon*, have been found in the Coal-formation of Nova Scotia. They are known to me only by the microscopic structure of their wood; but on the evidence afforded by this I have named and described them as new species*. One of them, *D. antiquius*, is closely allied to *D. Withami* of Great Britain, and, like that species, belongs to the Lower Carboniferous Coal-measures. Its structure is of that character for which Brongniart proposed the generic name "*Palæoxylon.*" It has not yet been found at the Joggins. Another species, *D. Acadianum*, is found abundantly at the Joggins in the condition of drifted trunks imbedded in the sandstone of the lower part of the Coal-formation and the upper part of the Millstone-grit series. The third species, *D. materiarium*, is very near to *D. Brandlingii* of Great Britain, and may possibly be only a variety. It is especially abundant in the sandstone of the Upper Coal-formation, in

* Descriptions referred to here and in subsequent pages will be found in "Synopsis of the Flora of the Carboniferous Period," Can. Nat. vol. viii., and in the Appendix.

which vast numbers of drifted trunks of this species occur in some
places. The fourth species, *D. annulatum*, presents a very peculiar
structure, probably of generic value. It has alternate concentric rings
of discigerous woody tissue, of the character of that of *Dadoxylon*, and
of compact structureless coal, which either represents layers of very
dense wood or, more likely, of corky cellular tissue. In the latter
case the structure would have affinities with that of certain *Gnetaceæ*
and of Cycads.

Though coniferous trees usually occur as decorticated and prostrate trunks, I have recorded the occurrence of one erect specimen,
in a sandstone a little above the "Main Coal," at the Joggins. It
probably belonged to the species last named. Tissues of coniferous
trees are very rare in the coal itself. Most of the discigerous tissues
found in the coal belong to *Sigillaria* and *Calamodendron*. From the
abundance of coniferous trees in sandstones above and below the coal,
and their comparative absence in the coal and coal-shales, it may be
inferred that these trees belonged rather to the uplands than to the
coal-swamps; and the great durability and small specific gravity of
coniferous wood would allow it to be drifted, either by rivers or ocean-
currents, to very great distances. I am not aware that the fruits of
pine trees occur at the Joggins, unless some of the *Trigonocarpa* are
of this character. Nor has any foliage of these trees been found;
but at Tatmagonche, in the continuation of the Upper Coal-formation,
there are leafy branchlets which I have named *Araucarites gracilis*,
and which may possibly have belonged to *Dadoxylon materiarium*.

The casts of pith-cylinders known as *Sternbergiæ* are abundant
in some of the sandstones, especially in the Upper Coal-formation.
I have shown that in Nova Scotia, as in England, some of these singular casts belong to *Dadoxylon*[*]; but as the pith-cylinder of *Sigillaria* and of *Lepidophloios* was of a similar character, those which are
destitute of woody investment cannot be determined with certainty,
though in general the transverse markings are more distant in the
Sternbergiæ of *Sigillaria* and *Lepidophloios* than in those of *Dadoxylon*.

In Plate V., and in Plate VI. fig. 14, I have given illustrations of
the coniferous plants above referred to.

3. *Sigillariæ.*—I have catalogued or described no less than twenty-
one species belonging to this family, from the Coal-measures of Nova
Scotia. They may be arranged under the following provisional
genera :—

(1.) FAVULARIA, *Sternberg* .... Sigillaria elegans, *Brongn.*
　　　　　　　　　　　　　　　—— tessellata, *Brongn.*
　　　　　　　　　　　　　　　—— Bretonensis, *Dawson.*
(2.) RHYTIDOLEPIS, *Sternberg* .. —— scutellata, *Brongn.*
　　　　　　　　　　　　　　　—— Schlotheimiana, *Brongn.*
　　　　　　　　　　　　　　　—— Saullii, *Brongn.*
　　　　　　　　　　　　　　　—— Dournaisii, *Brongn.*
　　　　　　　　　　　　　　　—— Knorrii, *Brongn.*

* Proceedings of the American Association, 1857, Canadian Naturalist, vol. ii.
Paper on Structures of Coal, Quart. Journ. Geol. Soc. 1860.

(2.) Rhytidolepis (*continued*)   Sigillaria pachyderma, *Brongn*.
                              —— flexuosa, *L. & H.*?
                              —— elongata, *Brongn*.
(3.) Sigillaria, *Brongn*.......   —— reniformis, *Brongn*.
                              —— Brownii, *Dawson*.
                              —— lævigata, *Brongn*.
                              —— planicosta, *Dawson*.
                              —— catenoides, *Dawson*.
                              —— striata, *Dawson*.
                              —— ominous, *Dawson*.
(4.) Clathraria, *Brongn*. ....   —— Menardi, *Brongn*.
(5.) Leioderma, *Goldenb*. ....   —— Sydnensis, *Dawson*.
    (Asolanus, *Wood*).
(6.) Syringodendron, *Sternb*...   —— organum, *L. & H.*

Of these, seven are probably new species, and the remainder can be identified with reasonable certainty with European species. The differences in the markings in different parts of the same tree are, however, so great, that I regard the greater part of the recognized species of *Sigillariæ* as merely provisional. Even the generic limits may be overpassed when species are determined from hand specimens. A fragment of the base of an old trunk of *Sigillaria* proper would necessarily be placed in the genus *Leioderma*, and a young branch of *Favularia* has all the characters of the genus *Clathraria*. It is, however, absolutely necessary to make some attempt at generic distinction among the diverse forms included in the genus *Sigillaria*; otherwise it will be impossible to reconcile the conflicting statements of authors as to the dimensions, habit of growth, foliage, roots, and fructification of these singular plants ;—such statements usually applying to one or more of the subordinate generic types. I shall therefore notice separately, and with especial reference to their function in the production of coal, the several generic or subgeneric forms, beginning with that which I regard as the most important—namely, *Sigillaria* proper, of which, in Nova Scotia, I regard the species which I have named *S. Brownii* (figs. 15 to 20, Pl. VI.) as the type. Other species are represented in figs. 21 to 28.

In the restricted genus *Sigillaria* the ribs are strongly developed, except at the base of the stem ; they are usually much broader than the oval or elliptical tripunctate areoles, and are striated longitudinally. The woody axis has both discigerous and scalariform tissues, arranged in wedges, with medullary rays as in exogens * ; the pith is transversely partitioned in the manner of *Sternbergia* ; and the inner bark contains great quantities of long and apparently very durable fibres, which I have, in my descriptions of the structures in in the coal, named "bast tissue." The outer bark was usually thick, of dense and almost indestructible cellular tissue. The trunk when old lost its regular ribs and scars, owing to expansion, and became furrowed like that of an old exogenous tree. The roots were *Stigmariæ* of the type of the ordinary *S. ficoides*. I have not seen the

---

* Quart. Journ. Geol. Soc., paper on Structures of Coal.

leaves or fruits attached; but, from the associations observed, I believe
that the former were long, narrow, rigid, and two- or three-nerved
(*Cyperites*), and that the latter were *Trigonocarpa*, borne in racemes
on the upper part of the stem. These trees attained to a great size.
I have seen one trunk four feet in diameter, and specimens of two
feet or more in diameter are common: some of these trunks have been
traced for thirty or forty feet without branching. The greater
number of the erect stumps preserved at the Joggins appear to belong
to this genus, which also seems to have contributed very largely to
the formation of coal. Judging from the paucity of their foliage, the
density of their tissues, and the strong structural resemblance of
their stems and roots to those of Cycads, I believe that their rate
of growth must have been very slow.

The genus *Rhytidolepis*, in which the areolæ are large, hexagonal,
and tripunctate, and the ribs narrow and often transversely striate,
ranks as a coal-producer next to *Sigillaria* proper, and is equally
abundant in the Coal-measures. These trees seem to have been of
smaller size and feebler structure than the last mentioned, and are
less frequently found in the erect position; but they are very abundant on the roofs of the coal-beds. Judging from such specimens
as I have seen, their roots were less distinctly Stigmarioid than in the
last genus, though this appearance may arise from difference of preservation. Their leaves were of the same type as in the last genus;
and their stems bear rings of irregular scars, which may mark stages
of growth, or the production of slender racemes of fruit in a verticillate manner. The woody axis of the stems of this genus was composed of scalariform and coarsely porous tissues, much like those of
modern Cycads. I figure, as an illustration of the genus, a fragment
of *S. scutellata* showing one of the belts of abnormal scars.

The genus *Favularia* is represented in Nova Scotia principally by
the typical species *S. elegans* of Brongniart. The admirable investiga-
tions of the structure of the stem of this species by Brongniart, with
the further illustrations given by Corda, Hooker, and Goldenberg, still
afford the best general views of the structure of *Sigillariæ* which
we possess [*]. It is to be observed, however, that Brongniart's specimen was a young stem or a branch, and that it affords a very imperfect idea of the development of discigerous and bast tissues in the
full-grown stems of *Sigillaria* proper. The trees of this genus appear to have been of small growth; and they branched in the manner
of *Lepidodendron*, the smaller branches being quite destitute of ribs,
and with the areolæ elliptical and spirally disposed. The stems show
joints or rings of peculiar scars at intervals, as in the last genus.
The leaves differ from those of the other genera, being broad and with
numerous slender parallel veins, almost in the manner of *Cordaites*.
(Figs. 26 and 27, Plate VII.)

The genus *Clathraria* is evidently closely allied to the above,
and is possibly founded on branches of trees of the genus *Favularia*.
It is a rare form in Nova Scotia.

---

* See also Binney " On some Fossil Plants showing Structure, from the Lower
Coal-measures of Lancashire," Quart. Journ. Geol. Soc. vol. xviii. p. 106.—EDIT.

Of the genus *Leioderma* or *Asolanus* I know but one species, independently of those specimens of old trunks of the ordinary *Sigillaria* in which the ribs have disappeared. My species, *S. Sydnensis*, is founded on specimens collected by Mr. Brown at Sydney, Cape Breton, which are especially remarkable for the curious modification which they present of the Stigmarian root. The specimens described by Mr. Brown under the name of *S. alternans* [*], and which have been copied by Geinitz and Goldenberg, belong, I believe, to this species. (Fig. 28, Plate VII.)

On the genus *Syringodendron* of Sternberg I have no observations to make. I have seen only fragments of stems; and these seem to be very rare.

I include under *Sigillariæ* the remarkable fossils known as *Stigmaria*, being fully convinced that all the varieties of these plants known to me are merely roots of *Sigillaria*; I have verified this fact in a great many instances, in addition to those so well described by Mr. Binney and Mr. Brown. The different varieties or species of *Stigmaria* are no doubt characteristic of different species of *Sigillaria*, though in very few cases has it proved possible to ascertain the varieties proper to the particular species of stem. The old view, that the *Stigmariæ* were independent aquatic plants, still apparently maintained by Goldenberg and some other palæobotanists, evidently proceeds from imperfect information. Independently of their ascertained connexion with *Sigillaria*, the organs attached to the branches are not leaves, but rootlets. This was made evident long ago by the microscopic sections published by Goeppert, and I have ascertained that the structure is quite similar to that of the thick fleshy rootlets of *Cycas*. The lumps or tubercles on these roots have been mistaken for fructification; and the rounded tops of stumps, truncated by the falling in of the bark or the compression of the empty shell left by the decay of the wood, have been mistaken for the natural termination of the stem [†]. The only question remaining in regard to these organs is that of their precise morphological place. Their large pith and regular areoles render them unlike true roots; and hence Lesquereux has proposed to regard them as *rhizomes*. But they certainly radiate from a central stem, and are not known to produce any true buds or secondary stems. In short, while their function is that of roots, they may be regarded, in a morphological point of view, as a peculiar sort of underground branches. They all ramify very regularly in a dichotomous manner, and, as Mr. Brown has shown, in some species at least, give off conical tap-roots from their underside.

In all the *Stigmariæ* exhibiting structure which I have examined, the axis exhibits only scalariform vessels. Corda, however, figures a species with wood-cells, or vessels with numerous pores, quite like those found in the stems of *Sigillaria* proper; and, as Hooker has pointed out, the arrangement of the tissues in *Stigmaria* is similar to that in

---

[*] Quart. Journ. Geol. Soc. vol. v. p. 354 *et seq.*

[†] For examples of the manner in which a natural termination may be simulated by the collapse of bark or by constriction owing to lateral pressure, see my papers, Quart. Journ. Geol. Soc. vol. x. p. 35, and vol. vii. p. 194.

Sigillaria. After making due allowance for differences of preservation, I have been able to recognize eleven species or forms of *Stigmaria* in Nova Scotia, corresponding, as I believe, to as many species of *Sigillaria**. At the Joggins, *Stigmariæ* are more abundant than any other fossil plants. This arises from their preservation in the numerous fossil soils or *Stigmaria* underclays. Their bark, and mineral charcoal derived from their axes, also abound throughout the thickness of the coal-beds, indicating the continued growth of *Sigillaria* in the accumulation of the coal. (Figs. 83 to 87 Pl. XII.)

Our knowledge of the fructification of *Sigillaria* is as yet of a very uncertain character. I am aware that Goldenberg has assigned to these plants leafy strobiles containing spore-capsules: but I do not think the evidence which he adduces conclusive as to their connexion with *Sigillaria*; and the organs themselves are so precisely similar to the strobiles of *Lepidophloios*, that I suspect they must belong to that or some allied genus. The leaves, also, with which they are associated in one of Goldenberg's figures seem more like those of *Lepidophloios* than those of *Sigillaria*. If, however, these are really the organs of fructification of any species of *Sigillaria*, I think it will be found that we have included in this genus, as in the old genus *Calamites*, two distinct groups of plants, one cryptogamous, and the other phænogamous, or else that male strobiles bearing pollen have been mistaken for spore-bearing organs.

I cannot pretend that I have found the fruit of *Sigillaria* attached to the parent stem; but I think that a reasonable probability can be established that some at least of the fruits included, somewhat vaguely, by authors under the names of *Trigonocarpum* and *Rhabdocarpus* were really fruits of *Sigillaria*. These fruits are excessively abundant and of many species, and they occur not only in the sandstones but in the fine shales and coals and in the interior of erect trees, showing that they were produced in the coal-swamps. The structures of these fruits show that they are phænogamous and probably gymnospermous. Now the only plants known to us in the coal-formation, whose structures entitle them to this rank, are the *Conifers*, *Sigillariæ*, and *Calamodendra*. All the others were in structure allied to cryptogams, and the fructification of most of them is known. But the Conifers were too infrequent in the Carboniferous swamps to have afforded numerous species of *Carpolites*; and, as I shall presently show, the *Calamodendra* were very closely allied to *Sigillariæ*, if not members of that family. Unless, therefore, these fruits belonged to *Sigillaria*, they must have been produced by some other trees of the coal-swamps, which, though very abundant and of numerous species, are as yet quite unknown to us. Some of the *Trigonocarpa* have been claimed for Conifers, and their resemblance to the fruits of *Salisburya* gives countenance to this claim; but the Conifers of the Coal-period are much too few to afford more than a fraction of the species. One species of *Rhabdocarpus* has been attributed by Geinitz to the genus *Nœggerathia*; but the leaves which he assigns to it are very like those of *Sigillaria elegans*, and may belong to some allied species. With

* See Appendix.

regard to the mode of attachment of those fruits, I have shown that one species, *Trigonocarpum racemosum* of the Devonian strata *, was borne on a rhachis in the manner of a loose spike, and I am convinced that some of the groups of inflorescence named *Antholithes* are simply young *Rhabdocarpi* or *Trigonocarpa* borne in a pinnate manner on a broad rhachis and subtended by a few scales.   Such spikes may be regarded as corresponding to a leaf with fruits borne on the edges, in the manner of the female flower of *Cycas*; and I believe with Goldenberg that these were borne in verticils at intervals on the stem.   In this case it is possible that the strobiles described by that author may be male organs of fructification containing, not spores, but pollen.   In conclusion, I would observe that I would not doubt the possibility that some of the fruits known as *Cardiocarpa* may have belonged to Sigillarioid trees.   I am aware that some so-called *Cardiocarpa* are spore-cases of *Lepidodendron*; but there are others which are manifestly winged nutlets allied to *Trigonocarpum*, and which must have belonged to phœnogams.   It would perhaps be unwise to insist very strongly on deductions from what may be called circumstantial evidence, as to the nature of the fruit of *Sigillaria*; but the indications pointing to the conclusions above stated are so numerous that I have much confidence that they will be vindicated by complete specimens, should these be obtained.   (Figs. 29 and 30, Pl. VII., and figs. 69 to 79, Pl. XII.)

All of the Joggins coals, except a few shaly beds, afford unequivocal evidence of *Stigmaria* in their underclays; and it was obviously the normal mode of growth of a coal-bed, that, a more or less damp soil being provided, a forest of *Sigillaria* should overspread this, and that the Stigmarian roots, the trunks of fallen *Sigillariæ*, their leaves and fruits, and the smaller plants which grew in their shade, should accumulate in a bed of vegetable matter to be subsequently converted into coal—the bark of *Sigillaria* and allied plants affording " bright coal," the wood and bast tissues mineral charcoal, and the herbaceous matter and mould dull coal.   The evidence of this afforded by microscopic structure I have endeavoured to illustrate in a former paper†.

The process did not commence, as some have supposed, by the growth of *Stigmaria* in ponds or lakes.   It was indeed precisely the reverse of this, the *Sigillaria* growing in a soil more or less swamp t not submerged, and the formation of coal being at last arrested y submergence.   I infer this from the circumstance that remains of cyprids, Fishes, and other aquatic animals are rarely found in the underclays and lower parts of the coal-beds, but very frequently in the roofs, while it is not unusual to find mineral charcoal more abundant in the lower layers of the coal.   For the formation of a bed of coal, the sinking and subsequent burial of an area previously dry seems to have been required.   There are a few cases at the Joggins where *Calamites* and even *Sigillariæ* seem to have grown on areas liable to frequent inundation; but in these cases coal did not accumulate.   The non-laminated, slickensided and bleached condition of most of the underclays indicates soils of considerable permanence.

* " Flora of the Devonian Period," Quart. Journ. Geol. Soc. vol. viii. p. 324.
† " On the Structures in Coal," Quart. Journ. Geol. Soc. 1859.

In regard to beds destitute of Stigmarian underclays, the very few cases of this kind apply only to shaly coals filled with drifted leaves, or to accumulations of vegetable mud capable of conversion into impure coal. The origin of these beds is the same with that of the carbonaceous shales and bituminous limestones already referred to. It will be observed in the section that in a few cases such beds have become sufficiently dry to constitute underclays, and that conditions of this kind have sometimes alternated with those favourable to the formation of true coal.

There are some beds at the Joggins, holding erect trees *in situ*, which show that *Sigillariæ* sometimes grew singly or in scattered clumps, either alone or amidst brakes of *Calamites*. In other instances they must have grown close together, and with a dense undergrowth of ferns and *Cordaites*, forming an almost impenetrable mass of vegetation.

From the structure of *Sigillariæ* I infer that, like Cycads, they accumulated large quantities of starch, to be expended at intervals in more rapid growth, or in the production of abundant fructification. I adhere to the belief expressed in previous papers that Brongniart is correct in regarding the *Sigillariæ* as botanically allied to the *Cycadaceæ*, and have recently more fully satisfied myself on this point by comparisons of their tissues with those of *Cycas revoluta*. It is probable, however, that when better known they will be found to have a wider range of structure and affinities than we now suppose.

There are some reasons for believing that the trees described by Corda under the names of *Diploxylon*, *Myelopithys*, and *Heterangium*, and also the *Anabathra* of Witham, are *Sigillariæ*. Much of the tissue discribed by Goeppert as *Araucarites carbonarius* is probably also Sigillarian.

4. *Calamodendron.*—The plants of this genus are quite distinct from *Calamites* proper. A *Calamodendron* as usually seen is a striated cast with frequent cross lines or joints; but when the whole stem is preserved, it is seen that this cast represents merely an internal pith-cylinder, surrounded by a woody cylinder composed in part of scalariform or reticulated vessels, and in part of wood-cells with one row of large pores on each side. External to the wood was a cellular bark, and the outer surface seems to have been simply ribbed in the manner of *Sigillaria*. It so happens that the internal cast of the pith of *Calamodendron*, which is really of the nature of a *Sternbergia*, so closely resembles the external appearance of the true Calamites as to be constantly mistaken for them. Most of these pith-cylinders of *Calamodendron* have been grouped in the species *Calamites approximatus*; but that species, as understood by some authors, appears also to include true *Calamites*[*], which, however, when well preserved, can always be distinguished by the scars of the leaves or branchlets which were attached to the nodes.

Calamodendron would seem, from its structure, to have been closely allied to *Sigillaria*, though, according to Unger, the tissues were dif-

* See Geinitz, "Steinkohlenformation in Sachsen."

ferently arranged, and the woody cylinder must have been much
thicker in proportion.

The tissues of *Calamodendron* are by no means infrequent in the
coal, and casts of the pith are common in the sandstones; but its
foliage and fruit are unknown. (Fig. 31, Pl. VII., *a* to *c*.)

5. *Calamites.*—Nine species of true *Calamites* have been recognized
in Nova Scotia, of which seven occur at the Joggins, the most abundant
being *C. Suckovii* and *C. Cistii.* The Calamites grew in dense brakes
on sandy and muddy flats. They were unquestionably allied to *Equi-
setaceæ*, and produced at their nodes either verticillate simple linear
leaves, as in *C. Cistii*, or verticillate branchlets with pinnate or verticillate leaflets, as in *C. Suckovii* and *C. nodosus.* The Calamites do not
seem to have contributed much to the growth of coal, though their
remains are not infrequent in it. The soils in which they most frequently grow were apparently too wet and liable to inundation and
silting up to be favourable to coal-accumulation. I have elsewhere
shown that some of the species of *Calamites* gave off numerous adventitious roots from the lower parts of their stems, and also multi-
plied by budding at their bases[*].

Of the genus *Equisetites* one species has been found in Cape Breton;
but it has not as yet been recognized at the Joggins. (Fig. 88, Pl.
XII.)

6. *Asterophyllites*, &c.—Five species of *Asterophyllites*, one of *Annularia*, five of *Sphenophyllum*, and three of *Pinnularia*, have been
found in Nova Scotia. I place these together as probably allied
plants. The *Pinnulariæ* were apparently slender roots, with thin
epidermis, cellular bark, and a central axis. The others were probably low plants growing in wet places. I am not aware that
they contributed to any great extent to the accumulation of coal; but
as their tissues were scalariform, similar to those of ferns, it would
not be easy to recognize them. A beautiful specimen of *Sphenophyllum emarginatum* from New Brunswick, in the collection of Sir
W. E. Logan, has enabled me to ascertain that its stem had a simple
axis of one bundle of reticulato-scalariform vessels, like those of *Tmesipteris* as figured by Brongniart. These curious plants were no
doubt cryptogamous, having a habit of growth like that of *Equise-
taceæ*, leaves like those of ferns or *Marsiliaceæ*, and fructification-
and structure like those of *Lycopodiaceæ.* They were closely allied
to *Asterophyllites* and *Annularia.*

7. *Filices.*—Of the numerous species of ferns in the Carboniferous
rocks of Nova Scotia, only a very few have been recognized at the Joggins. This may in part be due to the soft and crumbling character
of the shales; but after much examination I am inclined to believe that
the flora of the Joggins was originally poor in ferns. While the
coal-field of Sydney, Cape Breton, has afforded forty-six species,
the Joggins and its vicinity have as yet yielded only six or seven.
Of these by far the most abundant is *Alethopteris lonchitica*, which appears throughout the Middle Coal-formation under a great number
of varietal forms. It is also found abundantly at Springhill. At

* Quart. Journ. Geol. Soc. vol. x. p. 34

Sydney, on the contrary, *Pecopteris abbreviata* and *Alethopteris nervosa* are the most common ferns. But though fronds of ferns are comparatively rare at the Joggins, except in a few beds, and these holding principally the species *Alethopteris lonchitica*, bundles of scalariform vessels referable to ferns occur very plentifully in the coarser parts of the coal-beds, and would seem to indicate that vast quantities of stipes and fronds have been resolved into coal. It is to be observed, however, that it is not in all cases possible to distinguish the vascular bundles of ferns from those of the leaves of Sigillarioid and Lycopodiaceous plants. (Fig. 67, Pl. XII.)

Trunks of two species of tree ferns of the genus *Palæopteris* have been found in Nova Scotia and New Brunswick, and also obscure fragments, probably of *Caulopteris* and *Psaronius*. (Figs. 35 & 36, Pl. VIII.)

8. *Megaphyton.*—These are perhaps the most curious and puzzling plants of the coal. Their thick stems, marked by linear scars and having two rows of large depressed areoles on the sides, suggest no affinities to any known plants. They are usually ranked with *Lepidodendron* and *Ulodendron*, but sometimes, and probably with greater reason, are regarded as allied to tree ferns. At the Joggins a very fine species (*M. magnificum*) has been found, and at Sydney a smaller species (*M. humile*); but both are rare and not well preserved. If the large scars supported cones and the smaller leaves, then, as Brongniart remarks, the plant would much resemble *Lepidophloios*, in which the cone-scars are thus sometimes distichous. But the scars are not round and marked with radiating scales as in *Lepidophloios*; they are reniform or oval, and resemble those of tree ferns, for which reason they may be regarded as more probably leaf-scars*; and in that case the smaller linear scars would indicate ramenta, or small aërial roots. Further, the plant described by Corda as *Zippea disticha* is evidently a *Megaphyton*, and the structure of that species is plainly that of a tree fern of somewhat peculiar type. On these grounds I incline to the opinion of Geinitz, that these curious trees were allied to ferns, and bore two rows of large fronds, the trunks being covered with coarse hairs or small aërial roots. At one time I was disposed to suspect that they may have crept along the ground; but a specimen from Sydney shows the leaf-stalks proceeding from the stem at an angle so acute that the stem must, I think, have been erect. From the appearance of the scars it is probable that only a pair of fronds were borne at one time at the top of the stem; and if these were broad and spreading, it would be a very graceful plant. To what extent plants of this type contributed to the accumulation of coal I have no means of ascertaining, their tissues in the state of coal not being distinguishable from those of ferns and Lycopodiaceæ.

The species *Megaphyton humile* had, like Corda's *Zippea disticha*, a thick central axis striated longitudinally, and giving off very thick bundles of fibres, and probably scalariform vessels, to the bases of the leaves. (Figs. 33 &34, Pl. VIII.)

* This is the view of Lindley, 'Fossil Flora,' p. 116.

9. *Lepidodendron.*—Of this genus nineteen species have been recorded as occurring in the Carboniferous rocks of Nova Scotia. Of
these, six occur at the Joggins, where specimens of this genus are very
much less abundant than those of *Sigillaria.* In the newer Coal-
formation *Lepidodendra* are particularly rare, and *L. undulatum* is
the most common species. In the Middle Coal-formation *L. rimosum, L. dichotomum, L. cl. ens,* and *L. Pictoense* are probably the
most common species; and *L. corrugatum* is the characteristic *Lepidodendron* of the Lower Carboniferous, in which plants of this species
seem to be more abundant than any other vegetable remains whatever.
    To the natural history of this well-known genus I have little to
add, except in relation to the changes which take place in its trunk
in the process of growth, and the study of which is important in
order to prevent the undue multiplication of species. These are of
three kinds. In some species the areoles, at first close together,
become, in the process of the expansion of the stem, separated by intervening spaces of bark in a perfectly regular manner; so that in
old stems, while widely separated, they still retain their arrangement, while in young stems they are quite close to one another.
This is the case in *L. corrugatum* (Pl. XI.). In other species the
leaf-scars or areoles increase in size in the old stems, still retaining
their forms and their contiguity to each other. This is the case in
L. undulatum, and generally in those *Lepidodendra* which have very
large areoles. In these species the continued vitality of the bark is
shown by the occasional production of lateral strobiles on large
branches, in the manner of the modern Red Pine of America. In
other species the areoles neither increase in size nor become regularly
separated by growth of the intervening bark; but in old stems the
bark splits into deep furrows, between which may be seen portions
of bark still retaining the areoles in their original dimensions and
arrangement. This is the case with *L. Pictoense.* This cracking of
the bark no doubt occurs in very old trunks of the first two types,
but not at all to the same extent. I figure three examples of these
peculiarities in mode of growth:—
    Lepidodendron corrugatum, Dawson.—I quote in the Appendix
my description of this species, and may refer to the figures in Plate
XI. for further illustration. I do not know any other species in
Nova Scotia which has the same habit of growth; but *L. oculatum*
and *L. distans* of Lesquereux show a tendency to it. The present
species is exclusively Lower Carboniferous, and occurs on that
horizon in New Brunswick, in Pennsylvania, and, I believe, also in
Ohio; though the beds holding it in the latter State have been by
some regarded as Devonian.
    L. undulatum, Sternberg.—I think it not improbable that several
closely allied species are included under this name. On the other
hand, all the large-areoled *Lepidodendra* figured in the books must
have branches with small scars, which, in the present state of knowledge, it is impossible to identify with this species. I suppose that *L.
elegans* resembles the present species in its mode of growth, at least
if the large-scarred specimens attributed to it are really of the same

species. *L. dichotomum* (= *L. Sternbergii*) also resembles it to some extent. (Fig. 41, Pl. IX.)

L. Pictoense, Dawson.—This species I described as follows, from young stems, in my "Synopsis of the Coal-Plants of Nova Scotia:"— "Areoles contiguous, prominent, separated in young stems by a narrow line, long-oval, acuminate; breadth to length as 1 to 3 or less; lower half obliquely wrinkled, especially at one side. Middle line indistinct. Leaf-scar at upper end of areole, small, triangular, with traces of three vascular points, nearly confluent. Length of areole about 0·5 inch."

Additional specimens from Sydney show that in old trunks of this species the areoles do not enlarge, but the bark becomes split into strips. I have reason to think that a new species from Nova Scotia which I have described in the Appendix, *L. personatum*, agrees with it in this respect. (Figs. 37, 38, and 39. Pl. IX.)

The *Lepidodendra* resemble each other too closely to admit of good subgeneric distinction. The grounds on which the distinction of *Sagenaria* and *Aspidiaria* is founded are quite worthless, the apparent position of the vascular scars in the areoles depending on accidents of preservation much more than on original differences. The genus *Knorria* includes many peculiar conditions of decorticated *Lepidodendra*.

In regard to the accumulation of coal, *Lepidodendra*, when present, appear under the same conditions with *Sigillariæ*, the outer bark being converted into shining coal, and the scalariform axis appearing as mineral charcoal of a more loose and powdery quality than that derived from *Sigillaria*. On the planes of lamination of the coal the furrowed bark of old trunks can scarcely be distinguished from that of old *Sigillariæ*.

10. *Lepidophloios*.—Under this generic name, established by Sternberg, I propose to include those Lycopodiaceous trees of the Coal-measures which have thick ¹ .anches, transversely elongated leaf-scars, each with three vascular points and placed on elevated or scale-like protuberances, long one-nerved leaves, and large lateral strobiles in vertical rows or spirally disposed. Their structure resembles that of *Lepidodendron*, consisting of a *Sternbergia* pith, a slender axis of large scalariform vessels, giving off from its surface bundles of smaller vessels to the leaves, a very thick cellular bark, and a thin dense outer bark, having some elongated cells or bast tissue on its inner side.

Regarding *L. laricinum* of Sternberg as the type of the genus, and taking in connexion with this the species described by Goldenberg, and my own observations on numerous specimens found in Nova Scotia, I have no doubt that *Lomatophloios crassicaulis* of Corda, and other species of that genus described by Goldenberg, *Ulodendron* and *Bothrodendron* of Lindley, *Lepidodendron ornatissimum* of Brongniart, and *Halonia punctata* of Geinitz all belong to this genus, and differ from each other only in conditions of growth and preservation. Several of the species of *Lepidostrobus* and *Lepidophyllum* also belong to *Lepidophloios*.

The species of *Lepidophloios* are readily distinguished from *Lepidodendron* by the form of the areoles, and by the round scars on the stem, which usually mark the insertion of the strobiles, though in barren stems they may also have produced branches; still the fact of my finding the strobiles *in situ* in one instance, the accurate resemblance which the scars bear to those left by the cones of the Red Pine when borne on thick branches, and the actual impressions of the radiating scales in some specimens, leave no doubt in my mind that they are usually the marks of cones; and the great size of the cones of *Lepidophloios* accords with this conclusion.

The species of *Lepidophloios* are numerous, and individuals are quite abundant in the Coal-formation, especially toward its upper part. Their flattened bark is frequent in the coal-beds and their roofs, affording a thin layer of pure coal, which sometimes shows the peculiar laminated or scaly character of the bark when other characters are almost entirely obliterated. The leaves also are nearly as abundant as those of *Sigillaria* in the coal-shales. They can readily be distinguished by their strong angular midrib.

I figure, in illustration of the genus, all the parts known to me of *L. Acadianus*, and characteristic specimens of other species. One of these, *L. parvus*, is characteristic of the Upper Coal-formation. (*Vide* Plate X. & Plate XI. fig. 51.)

11. *Cordaites* or *Pychnophyllum.*—This plant is represented in the Coal-formation chiefly by its broad striated leaves, which are extremely abundant in the coal and its associated shales. Some thin coals are indeed almost entirely composed of them. The most common species is *C. borassifolia*, a plant which Corda has shown to have a simple stem with a slender axis of scalariform vessels resembling that of *Lepidophloios*; for this reason, notwithstanding the broad and parallel-veined leaves, I regard this genus as belonging to *Lycopodiaceæ* or some allied family. It must have been extremely abundant in the Carboniferous swamps; and, from the frequency of its being covered with *Spirorbis*, I think it must either have been of more aquatic habit than most of the other plants of the Coal-formation, or that its leaves must have been very durable. While the leaves are abundant, the stems are very rare. I infer that they were usually low and succulent. Much of the tissue found in the coal, which I have called "epidermal," probably belongs to leaves of *Cordaites.*

In the Upper Coal-formation there is a second species, distinguished by its simple and uniform venation. This I have named *C. simplex.*

12. *Sporangites.*—To avoid the confusion which envelopes the classification of Carpolites, I have used the above name for rounded spore-cases of *Lepidodendron* and allied plants, which are very frequent in the coal. A smooth round species like a mustard-seed, is excessively abundant in the Lower Carboniferous at Horton, and probably belongs to *Lepidodendron corrugatum*, with which it is associated. A species covered with papillæ, *S. papillata*, constitutes nearly the whole of some layers in coal 12, group xix. of the preceding Section.

I have no indication as to the plant to which it may belong, except that it is associated with *Cordaites*. (Figs. 80 & 81, Pl. XII.)[*]

13. *Tissues in the Mineral Charcoal.*—On these I have little to add to the statements in my paper of 1859, "On the Vegetable Structures in Coal"[†]. These tissues may be arranged as follows:—

a. *Bast tissue, or elongated cells from the liber or inner bark of Sigillariæ and Lepidodendron, but especially of the former.*—This kind of tissue is abundant in a calcified state in the shales associated with the coals, and also as mineral charcoal in the coals themselves, and in the interior of erect *Sigillariæ*. It is the kind of tissue figured by Brongniart as the inner layer of bark in *Sigillaria elegans*, and very well described by Binney (Quart. Journ. Geol. Soc. vol. xviii.) as " clongated tissue or utricles." Under the microscope many specimens of it closely resemble the imperfect bast tissue of the inner bark of *Pinus strobus* and *Thuja occidentalis*; and like this it seems to have been at once tough and durable, remaining in fibrous strips after the woody tissues had decayed. It is extremely abundant at the Joggins in the mineral charcoal of the smaller coal-seams. It is often associated with films of structureless coal, which represent the dense cellular outer bark which, in the trunk of *Sigillaria*, not only surrounded this tissue, but was intermixed with it. (Fig. 62, Pl. XII.)

b. *Vascular bundles of Ferns.*—These may be noticed by all close observers of the surfaces of coal, as slender hair-like fibres, sometimes lying separately, in other cases grouped in bands half an inch or more in diameter, and imbedded in a loose sort of mineral charcoal. When treated with nitric acid, each bundle resolves itself into a few scalariform vessels surrounded with a sheath of woody fibres, often minutely porous. This structure is precisely that of macerated fern-stipes; but, as already stated, there may have been some other coal-plants whose leaves presented similar bundles. As stated in my former paper " On the Vegetable Structures in Coal," this kind of tissue is especially abundant in the coarse and laminated portions of the coal, which we know on other evidence to have been made up, not of trunks of trees, but of mixed herbaceous matters. (Pl. XII. fig. 67.)

c. *Scalariform vessels.*—These are very abundant in the mineral charcoal, though the coarser kinds have been crushed and broken in such a manner that they usually appear as mere débris. The scalariform vessels of *Lepidodendron*, *Lepidophloios*, and *Stigmaria* are very coarse and much resemble each other. Those of ferns are finer, and sometimes have a reticulated structure. Those of *Sigillaria* are much finer and often have the aspect of wood-cells with transversely elongated pores like those of *Cycas*. Good examples of these are figured in the paper already referred to. (See also Plate XI. figs. 54, &c.)

d. *Discigerous wood-cells.*—These are the true bordered pores

* It much resembles the spore-cases of *Flemingites gracilis*, as figured by Carruthers, 'Geol. Mag.' vol. ii. I suppose this to be a strobile of *Lepidophloios*.
† Quart. Journ. Geol. Soc. February 1860.

characteristic of *Sigillaria, Calamodendron,* and *Dadoxylon.* In the two former genera the disks or pores are large and irregularly arranged, either in one row or several rows. In the latter case they are sometimes regularly alternate and contiguous. In the genus *Dadoxylon* they are of smaller size and always regularly contiguous in two or more rows, so as to present an hexagonal areolation. Discigerous structures of *Sigillaria* and *Calamodendron* are very abundant in the coal, and numerous examples were figured in my former paper. I have indicated by the name *Reticulated Tissue* certain cells or vessels which may either be reticulated scalariform vessels, or an imperfect form of discigerous tissue. I believe them to belong to *Stigmaria* or *Calamodendron.* (Figs. 57 & 68, Pl. XII.)

o. *Epidermal tissue.*—This is a dense cellular tissue representing the outer integuments of various leaves, herbaceous stems, and fruits. I have ascertained that the structures in question occur in the leaves and stipes of *Cordaites* and ferns, and in the outer coat of *Carpolites* and *Sporangites.* With this I may include the obscure and thick-walled cellular tissue of the outer bark of *Sigillaria* and *Lepidodendron* and other trees, which, though usually consolidated into compact coal, sometimes exhibits its structure.

I would here emphatically state that all my observations at the Joggins confirm the conclusion, which I arrived at many years ago from the study of the coals of Pictou and Sydney, that tho layers of clear shining coal (pitch or cherry coal) are composed of flattened trunks of trees, and that of these usually the bark alone remains; further that the lamination of the coal is due to the superposition of layers of such flattened trunks alternating with the accumulations of vegetable matter of successive years, and occasionally with fine vegetable muck or mud spread over the surface by rains or by inundations. In connexion with this, it is to be observed that the density and *impermeability* of cortical tissues not only enable them to endure after wood has perished or been resolved into bits of charcoal, but render them less liable than the wood to mineral *infiltration.*

14. *Rate of Growth of Carboniferous Plants.*—Very vague statements are often made as to the supposed rapid rate of growth of plants in the Carboniferous period. Perhaps the most trustworthy facts in relation to this subject are those which may be obtained from the coniferous trees. In some of these (for instance, *Dadoxylon materiarium, D. annulatum,* and *D. antiquus*) the rings of growth, which were no doubt annual, are distinctly marked. On measuring these in a number of specimens, and comparing them with modern species, I find that they are about equal in dimensions to those of the Balsam-Fir or the Yellow Pine of America. Assuming, therefore, similarity in habit of growth and extent of foliage to these species, we may infer that, in regard to coniferous trees, the ordinary conditions of growth were not dissimilar from those of Eastern America in its temperate regions at present. When, however, we compare the ferns and *Lycopodiaceæ* of the Coal-formation with those now growing in Eastern America, we see, in the much greater dimensions and luxuriance of the former, evidence of a much

more moist and equable climate than that which now subsists; so that we may suppose the growth of such plants to have been more rapid than it is at present. These plants would thus lead us to infer a warm and insular climate, perhaps influenced by that supposed excess of carbonic acid in the atmosphere, which, as Tyndall and Hunt inform us, would promote warmth and moisture by impeding terrestrial radiation. With this would also agree the fact that the Conifers have woody tissues resembling those of the pine trees of the milder climates of the southern hemisphere at present.

If we apply these considerations to *Sigillaria*, we may infer that the conditions of moisture and uniformity of temperature favourable to ferns and *Lycopodiaceæ* were also favourable to these curious plants. They must have been perennial; and the resemblance of their trunks to those of Cycads, together with their hard and narrow leaves, would lead us to infer that their growth must have been very slow. A similar inference may be drawn from the evidences of very slow and regular expansion presented by the lower parts of their stems. On the other hand, the distance, of a foot or more, which often intervenes between the transverse rows of scars, marking probably annual fructification, would indicate a more rapid rate of growth. Further, it may be inferred, from the structure of their roots and of their thick inner bark, that these, as in Cycads, were receptacles for great quantities of starch, and that the lives of these plants presented alternations of starch-accumulation and of expenditure of this in the production of leaves, wood, and abundant inflorescence. They would thus, perhaps for several years, grow very slowly, and then put forth a great mass of fructification, after which perhaps many of the individuals would die, or again remain for a long time in an inactive state. This view would, I think, very well harmonize with the structure of these plants, and also with the mode of their entombment in the coal.

From the manner of the association of Calamites with erect *Sigillariæ*, I infer that the former were, of all the plants of the Coal-formation, those of most rapid dissemination and growth. They appear to have first taken possession of emerging banks of sand and mud, to have promoted the accumulation of sediment on inundated areas, and to have protected the exposed margins of the forests of *Sigillariæ*.

In applying any conclusions as to the rate of growth of Carboniferous plants to the accumulation of coal, we must take into account the probable rate of decay of vegetable matter. When we consider the probable wetness of the soils on which the plants which produced the coal grew, the density of the forests, and the possible excess of carbonic acid in the atmosphere of these swamps, we must be prepared to admit that, notwithstanding the warmth and humidity, the conditions must have been favourable to the preservation of vegetable matter. Still the hollow cylinders of bark, the little fragments of decayed wood in the form of mineral charcoal, and the detached vascular bundles of ferns, testify to an enormous amount of decay, and show that, however great the accumulation of coal, it reprosents only a fraction of the vegetablo matter which was actually produced. It has been estimated that it would require eight feet of compact vegetable matter to produce one foot of coal*; but if we reckon the whole vegetable matter actually produced in the process, I should suppose that five times that amount would bo far below the truth, even in the most favourable cases; while there is evidence that in the Carboniferous period many forests may have flourished for centuries without producing an inch of coaly matter.

15. *Bivalve Shells.*—All the Lamellibranchiate shells, which are so numerous in some of the shales and bituminous limestones of the Joggins that some of the beds may bo regarded as composed of them, belong to one generic or family group. They are the so-called Modiolas, Unios, or Anodons of authors. I proposed for them some years ago the generic name of *Naiadites*, and described six species from the Coal-measures of Nova Scotia, stating my belief that they are allied to *Unionidæ*, and that their nearest analogue may bo the genus *Bysso-anodonta* of D'Orbigny, found in the River Papanat†. Mr. Salter, however, to whom I sent specimens, regards these shells as belonging to his new genera *Anthracomya* and *Anthracoptera*, the former being supposed to be allied to *Myadæ‡*. More recently Gümbel and Geinitz have described similar shells from Thuringia as belonging to the genera *Unio* and *Anodon*, and regard my *Naiadites carbonarius* (*Anthracoptera carbonaria* of Salter) as a *Dreissena*§. As these shells swarmed in the waters of the Coal-formation estuaries or lagoons, facts tending to the elucidation of their habits and affinities are important with reference to the coal; I would therefore make the following remarks in relation to them:—

(1) Under the microscope, the shells of the thicker species, as *Naiadites carbonarius*, present an internal lamellar and subnacreous layer, a thin layer of vertical prismatic shell, and an epidermis—these structures being entirely similar to those of *Unionidæ*. In the thinner species, as in *N. lævis*, only the prismatic coat appears, and in this the prisms are in some instances placed obliquely. These thin shells, however, show evidence of an epidermis. (2) The ligament was external, there seem to have been no teeth, the shell was closed posteriorly; but there are indications of a byssal sinus. Mr. Salter describes the epidermis as wrinkled posteriorly; but this, with the exception of the rings of growth, appears to me to result from pressure. The shells are equivalve, and have the external aspect of *Unionidæ* or *Mytilidæ*. (3) I know of no instance in Nova Scotia of the occurrence of these shells in the strictly marine limestones, nor have any properly marine forms of Mollusca been found with *Naiadites* in the Coal-measures. (4) The mode of their occurrence precludes the idea that they were burrowers, but favours the belief that they were attached by a byssus to sunken or floating timber. On the whole I think that the balance of probability is in

* Dana's Manual, p. 367.
† Supplement to Acadian Geology, 1860.
‡ Quart. Journ. Geol. Soc. vol. xix. p. 79.
§ Neues Jahrbuch, 1864. Geological Magazine, May 1865.

favour of the conclusion that they were brackish-water or fresh-water shells, allied to *Mytilidæ* or to embryonic *Unionidæ*.

16. *Spirorbis carbonarius.*—This little shell, which I described as a *Spirorbis* as long ago as 1845[*], is apparently not specifically distinct from *Microconchus carbonarius* of the British coal-fields. Its micro-scopic structure is identical with that of modern *Spirorbes*, and shows that it is a true worm-shell. It is found throughout the Coal-formation, attached to plants and to shells of *Naiadites*, and must have been an inhabitant of enclosed lagoons and estuaries. Its occurrence on *Sigillariæ* has been used as an argument in favour of the opinion that these trees grew in sea-water; but, unfortunately for that conclusion, the *Spirorbis* is often found on the inside of *Sigil-laria*-bark, showing that this had become dead and hollow. Beside this, the same kind of evidence would prove that *Lepidodendra, Cor-daites*, and Ferns were marine plants. *Spirorbes* multiply fast and grow very rapidly; and these little shells no doubt took immediate possession of submerged vegetation, just as their modern allies cover fronds of *Laminaria* and *Fucus*.

As I have not met with a description of this little shell, I may state that it is dextral, with $2\frac{1}{2}$ to 3 turns. It is attached through-out its length, and when not compressed presents a somewhat deep umbilicus. It is closely marked with beaded or unequal transverse ridges. It has when young a close resemblance to *Sp. caperatus*, M'Coy, from the Carboniferous Limestone of Ireland; but this species has only two turns, and is sinistral.

17. *Crustacea.*—It appears in the table above that as many as four-teen beds of coal exhibit in their roofs shells of minute Entomostraca of the genera *Cythere* and *Bairdia*; and these occur in such quan-tities that considerable beds of shale and bituminous limestone are filled with their valves. Professor Jones regards the species as marine or brackish-water; and the same remark will, I presume, apply to the crustacean *Diplostylus Dawsoni*, and a fragment of *Eurypterus* described by Mr. Salter from Coal no. 8 of Division 4 of the Section. Of the small Entomostracans there are several species, which Professor Jones has now in his hands for determina-tion. No Estherians have yet been found in the Coal-formation of Nova Scotia; but I have specimens of *Leaia Leidyi* from the Lower Carboniferous of Plaister Cove, and an undetermined *Estheria* from the same horizon at Horton Bluff.

It is to be observed that *Naiadites, Spirorbis*, and *Cythere* con-stantly occur associated in the same beds; and the conclusions as to habitat applicable to any one of these genera must apply to all.

18. *Fishes.*—Remains of fishes occur in connexion with eighteen of the coal-beds at the Joggins, usually in the roof-shales, though detached scales, teeth, spines, or coprolites are of occasional though rare occurrence in the coal itself, especially where the latter passes into coarse coal or carbonaceous shale. One thin bed, no. 6 of Division 4 of the Section, is full of remains of small fishes. It is hard and laminated, and roofed with a calcareous bed full of remains

* Quart. Journ. Geol. Soc. vol. i. p. 326.

of aquatic animals. It has a true Stigmarian underclay. I suppose
it to have been a swamp or forest submerged and occupied by fishes
while its vegetation was still standing. It contains remains of fishes
of the genera *Ctenoptyzhius, Diplodus, Rhizodus,* and *Palœniscus.*
It also contains *Cythere, Naiadites,* and *Spirorbis.* In the other
beds which contain fish-remains, most of these consist of small Lepi-
doganoids, but there are occasional teeth and scales of large species
of *Rhizodus,* and also teeth and scales of Selachian fishes of considerable size.

19. *Land-animals.*—The Coal-formation of Nova Scotia has afforded
the remains of eight species of Reptiles or Batrachians, belonging to
the genera *Hylonomus, Baphetes, Dendrerpeton, Hylerpeton,* and *Eosaurus;* of one Myriapod, *Xylobius sigillariæ;* of one land snail, *Pupa
vetusta;* and of one insect. All of these, except one of the reptiles,
have been found at the Joggins. I have nothing in regard to them
to add to what I have already published in my Memoir on 'Air-
breathers of the Coal Period.'

V. APPENDIX.

*Descriptive List of Carboniferous Plants found in Nova Scotia and
New Brunswick.*

[Abridged and corrected from " Synopsis of the Carboniferous Flora of Nova
Scotia," Can. Nat. vol. viii.] *

DADOXYLON, Unger.

1. DADOXYLON ACADIANUM, spec. nov. Pl. V. figs. 4–6.

Large trees, usually silicified or calcified, with very wide wood-
cells, having three or more rows of small hexagonal areoles, each
enclosing an oval pore; cells of modullary rays one-third of breadth
of wood-cells, and consisting of twenty or more rows of cells superimposed in two series. Rings of growth indistinct.

M. C.†, Joggins, Port Hood, Dorchester (*J. W. D.*).

2. D. MATERIARIUM, spec. nov. Pl. V. figs. 7–9.

Wood-cells less wide than those of the last; two to rarely four
rows of hexagonal disks. Medullary rays very numerous, with
twenty or more rows of cells superimposed in one series. Rings of
growth slightly marked. Approaches in the character of its woody
fibre to *D. Brandlingii:* but the medullary rays are much longer.
Some specimens show a large Sternbergian pith, with transverse
partitions‡. Vast numbers of trunks of this species occur in some
sandstones of the Upper Coal-formation.

M. and U. C., Joggins, Malagash, Picton, &c. (*J. W. D.*); Glace
Bay (*H. Poole*); Miramichi (*G. F. Matthew*).

* The illustrations are principally from photographs by my son George M.
Dawson, and for the sake of economy have been confined to small and character-
istic portions of the specimens.
† U. C., M. C., and L. C., indicate the Upper, Middle, and Lower Coal-for-
mations.
‡ Canadian Naturalist, 1857.

3. D. ANTIQUUS, spec. nov.   Pl. V. figs. 1–3.

Wood-cells narrow, thick-walled, two to three rows of pores. Medullary rays of three or four series of cells with twenty or more superimposed, nearly as wide as the wood-cells. Rings of growth visible. This species would belong to the genus *Palaoxylon* of Brongniart, and is closely allied to *D. Withami*, L. and H., which, like it, occurs in the Lower Coal-measures.

L. C., Horton (*Dr. Harding*).

4. D. ANNULATUM, spec. nov.   Pl. V. figs. 10–13.

Wood-cells with two or three rows of hexagonal disks. Medullary rays of twenty or more rows of cells superimposed, in two series. Wood divided into distinct concentric circles, alternating with layers of structureless coal representing cellular tissue or very dense wood. A stem 6 inches in diameter has fourteen to sixteen of these rings, and a pyritized pith about 1 inch in diameter. This is probably generically distinct from the preceding species.

M. C., Joggins (*Sir W. E. Logan*; *J. W. D.*).

ARAUCARITES, Unger.

ARAUCARITES GRACILIS, spec. nov.   Pl. VI. fig. 14.

Branches slender, 0·2 inch in diameter, with scaly, broad leaf-bases. Branchlets pinnate, numerous, very slender, with small, acute, spirally disposed leaves.

U. C., Tatamagouche (*J. W. D.*).

SIGILLARIA, Brongn.

1. SIGILLARIA (FAVULARIA) ELEGANS, Brongn.   Pl. VII. fig. 26.

Abundant, especially in the roofs of coal-seams. *S. hexagona* includes old trunks of this species. Young branches have scars of an elliptical form like those of *S. Serlii*.

M. C., Joggins (*J. W. D.*); Sydney (*R. Brown*).

2. S. (FAV.) TESSELLATA, Brongn.

M. C., Joggins and Pictou (*J. W. D.*); Sydney (*R. Brown*).

3. S. (RHYTIDOLEPIS) SCUTELLATA, Brongn.   Pl. VI. fig. 25.

M. and U. C., Joggins (*Lyell*; *J. W. D.*).

4. S. (RH.) SCHLOTHEIMIANA, Brongn.

M. C., Joggins (*Lyell*; *J. W. D.*).

5. S. (RH.) SAULLII, Brongn.

M. C., Sydney (*R. Brown*); Joggins (*Lyell*; *J. W. D.*).

6. S. BROWNII, Dawson (Quart. Journ. Geol. Soc. vol. x.).   Pl. VI. figs. 15–19.

M. C., Joggins (*J. W. D.*).

7. S. RENIFORMIS, Brongn.

M. C., Joggins (*Lyell*; *J. W. D.*); Sydney (*R. Brown*).

8. S. LÆVIGATA, Brongn.

M. C., Sydney (*R. Brown*); Joggins (*J. W. D.*).

9. S. PLANICOSTA, spec. nov.   Pl. VI. fig. 21.

Scars half hexa.:onal above, rounded below; lateral vascular impressions elongate; central small, punctiform.   Ribs 1·1 inch broad, smooth externally, longitudinally striated on the ligneous surface. Slight transverse wrinkles between the scars, which are distant from each other about an inch.   Allied to *S. lævigata*, but with very thin bark.

M. C., Sydney (*R. Brown*).

10. S. CATENOIDES, spec. nov.   Pl. VI. fig. 22.

Cortical surface unknown; ligneous surface with puncto-striate ribs 1·1 inch in breadth, and with single oval scars half an inch long, and an inch distant from centre to centre.   A very large tree. Perhaps, if its cortical surface were known, it might prove to be a large *Syringodendron*.

M. C., Joggins (*J. Smith*); Sydney (*R. Brown*).

11. S. STRIATA, spec. nov.   Pl. VI. fig. 23.

Ribs prominent, coarsely striate, 0·35 inch wide.   Scars nearly as wide as the ribs, rounded, hexagonal, 1 inch distant; lateral vascular marks narrow, central large.   On the ligneous surface scars single, round, oblong; bark very thin.

M. C., Joggins (*J. W. D.*).

12. S. ——?

A small erect stem, somewhat like *S. flexuosa*.

M. C., Joggins (*J. W. D.*).

13. S. (CLATHRARIA) MENARDI, Brongn.

M. C., Sydney (*R. Brown*); U. C., Picton (*J. W. D.*).

14. S. (ASOLANUS) SYDNENSIS, spec.   ".   Pl. VII. fig. 28.

Ribs obsolete; cortical and ligneou   surfaces striate; vascular scars two, elongate longitudinally, and alik   on cortical and ligneou surfaces; scars 1·1 inch distant, in rows 0·6 in   istant.   Stigmarian roots, same with variety *h* of *Stigmaria*, as described below.

M. C., Sydney (*R. Brown*).

15. S. ORGANUM, L. & H.

M. C., Sydney (*R. Brown*).

16. S. ELONGATA, Brongn.

M. C., Sydney (*R. Brown*).

17. S. FLEXUOSA, L. & H.

M. C., Sydney (*R. Brown's list in 'Acadian Geology'*).

18. S. PACHYDERMA, L. & H.

M. C., Sydney (*R. Brown's list*).

19. S. (FAV.) BRETONENSIS, spec. nov. Pl. VII. fig. 27.

Like *S. tessellata*, but areoles more hexagonal, bark thin and smooth on both sides, and furrow above the scars arcuate and with a central punctiform elevation.

M. C., Sydney (*R. Brown*).

20. S. EMINENS, spec. nov. Pl. VI. fig. 24.

Like *S. obovata*, Lesqx., but with narrower ribs, and larger and less distant areoles, each with a slight groove above.

M. C., Sydney (*R. Brown*).

21. S. DOURNANSII, Brongn.

M. C., Joggins (*J. W. D.*).

22. S. KNORRII, Brongn.

M. C., Sydney (*R. Brown*).

SYRINGODENDRON, Brongn.

Obscure specimens, referable to a narrow-ribbed species of this genus, occur in the Lower Carboniferous beds at Horton and Onslow.

STIGMARIA, Brongn.

STIGMARIA FICOIDES, Brongn. Pl. XII. fig. 83–87.

Under this name I place all the roots of *Sigillariæ* occurring in the Carboniferous rocks of Nova Scotia. They belong, without doubt, to the different species of Sigillarioid trees; but it is at present impossible to determine to which; and the specific characters of the *Stigmariæ* themselves are, as might be anticipated, evanescent and unsatisfactory. The varieties which occur in Nova Scotia, discarding mere difference of preservation, may be arranged as follows:—

Var. *a*. Areoles large, distant; bark more or less smooth. This is the most common variety, and extends throughout the Coal-formation.

Var. *b*. Areoles large, separated by waving grooves of the bark.

Var. *c*. Similar, but ridges as well as furrows between the areoles; var. *undulata* of Goeppert.

Var. *d*. Areoles small, separated by waving grooves.

Var. *e*. Areoles moderate, in vertical or diagonal furrows separated by ridges; var. *sigillarioides* of Goeppert.

Var. *f*. Areoles small; bark finely netted with wrinkles or striæ.

Var. *g*. Areoles surrounded by radiating marks, giving a star-like form; var. *stellata* of Goeppert. The only specimen I have seen was found by Dr. Harding in the Lower Carboniferous Coal-measures of Horton.

Var. *h*. Areoles small, or obscure and infrequent. Surface covered with fine uneven striæ. My specimens were collected by Mr. Brown in the Middle Coal-measures at Sydney.

Var. *i*. Areoles narrow, elongate, bark smooth or striate.

Var. *k*. *alternans*, with areoles in double rows on broad ribs separated by deep furrows. Probably old furrowed roots.

Var. l. Knorroides. Prominent bosses or ridges instead of areoles. These are imperfectly preserved specimens.

The varieties *a, b, c, e, i,* have been seen attached to trunks of *Sigillariæ* of the group distinguished by broad and prominent ribs—*Sigillaria* proper of the above arrangement. *Stigmariæ,* like *Sigillariæ,* are exceedingly abundant in the Middle Coal-measures, and are comparatively rare in the Lower Carboniferous and newer Coal-formations.

CALAMODENDRON, Brongn.

1. CALAMODENDEON APPROXIMATUM, Brongn. Pl. VII. fig. 31.

This plant is evidently quite distinct from *Calamites* proper. The Calamite-like cast is a pith or internal cavity, surrounded by a thick cylinder of woody tissue consisting of scalariform vessels and woody fibres with one row of round pores; external to this is a bark of cellular and bast tissue. The structure appears to be allied to that of *Sigillaria,* and is one of the most common in the beds of bituminous coal.

M. C., Sydney (*R. Brown*); M. C., Joggins, Pictou (*J. W. D.*); Coal Creek (*C. B. Matthew*).

2. C. OBSCURUM, spec. nov. Pl. VII. fig. 31 *d.*

This is a Calamite-like fragment found in a block of Sydney coal, in the state of mineral charcoal. The external markings are obscure, but the structure is well preserved. It differs from the last in having large ducts with many rows of pores, or reticulated instead of scalariform vessels. This is perhaps a Calamite.

M. C., Sydney (*J. W. D.*).

CYPERITES, L. & H.

CYPERITES —— ?

These elongate linear leaves have two or three ribs, and the central band between the ribs raised above the margin; one species has been seen attached to *Sigillaria Schlotheimiana.*

The leaves of *Sigillaria elegans* are different, being as broad as the areoles of the stem, and with several parallel veins.

Middle and upper coals, everywhere.

ANTHOLITHES, Brongn.

1. ANTHOLITHES RHABDOCARPI, spec. nov. Pl. VII. fig. 30.

Stem short, interruptedly striate, with two rows of crowded ovate fruits, and traces of floral leaves. Fruits half an inch long, striated longitudinally, attached by short peduncles.

M. C., Grand Lake (*C. F. Hartt*).

2. A. PYGMÆA, spec. nov. Pl. VII. fig. 30 *c.*

Rhachis 1 inch thick, rugose; two rows of opposite flowers, each showing four lanceolate striate floral leaves, two outer and two inner.

M. C., Joggins (*J. W. D.*).

3. A. SQUAMOSA, spec. nov.   Pl. VII. fig. 29.

Rhachis thick, coarsely rugose, with two rows of closely placed cones or scaly fruits.

U. C., Pictou (*J. W. D.*).

4. A. ——?, spec. nov.

Indistinct, but apparently different from those above described.

M. C., Joggins (*J. W. D.*); Sydney (*R. Brown*).

TRIGONOCARPUM, Brongn.

1. TRIGONOCARPUM HOOKERI, Dawson, Quart. Journ. Geol. Soc. vol. xvii.

M. C., Mabou (*J. W. D.*).

2. T. SIGILLARIÆ, spec. nov.   Pl. XII. fig. 76.

Ovate, ¼ inch long; testa smooth, or rugose longitudinally, acuminate, two-edged.   Found in erect trunks of *Sigillariæ* in large numbers.

M. C., Joggins (*J. W. D.*).

3. T. INTERMEDIUM, spec. nov.   Pl. XII. fig. 78.

Allied to *T. olivæformis*, but larger and more elongated.

M. C., Joggins, (*J. W. D.*).

4. T. AVELLANUM, spec. nov.   Pl. XII. fig. 77.

Allied to *T. ovatum*, L. & H.; three-ribbed, size and form of a filbert.

M. C., Joggins (*J. W. D.*); Sydney (*R. Brown*).

5. T. MINUS, spec. nov.   Pl. XII. fig. 75.

Half the size of *T. Hookeri*, and similar in form.

M. C., Joggins (*J. W. D.*).

6. T. ROTUNDUM, spec. nov.

Small, round-ovate, slightly pointed.

M. C., Joggins (*J. W. D.*).

7. T. NŒGGERATHII, Brongn.

Newer Coal-formation, Pictou (*J. W. D.*).

RHABDOCARPUS, Goepp. and Bergm.

1. RHABDOCARPUS ——?, spec. nov.

Ovate acuminate, less than half an inch long.

M. C., Joggins (*J. W. D.*).

2. R. INSIGNIS, spec. nov.   Pl. XII. fig. 89.

1·5 inch long, ovate, smooth, with about seven ribs on one side, and the intervening surface obscurely striate.   The nature of this fossil is perhaps doubtful; but if a fruit, it is the largest I have seen in the Coal-formation.

U. C., Pictou (*J. W. D.*).

CALAMITES, Suckow.

1. CALAMITES SUCKOVII, Brongn.

This species is one of the most common in an erect position. It has verticillate branchlets, with pinnate linear leaflets.

M. C., Sydney (*R. Brown*); Joggins (*Lyell; J. W.D.*); Grand Lake (*C. F. Hartt*); U. C., Pictou (*J. W. D.*); Coal Creek (*C. B. Matthew*).

2. C. CISTII, Brongn.

M. C., Joggins (*J. W. D.*); Sydney (*R. Brown*); Grand Lake (*C. F. Hartt*); Bay de Chaleur (*Logan*); Coal Creek (*C. B. Matthew*).

Often found erect. Its leaves are verticillate, simple, linear, striate, apparently one-nerved, and 3 inches long.

3. C. CANNÆFORMIS, Brongn.

M. C., Joggins (*Lyell; J. W. D.*); Sydney (*R. Brown*).

4. C. RAMOSUS, Artis.

Possibly a variety of *C. Suckovii*.
M. C., Joggins (*J. W. D.*); Sydney (*R. Brown*).

5. C. VOLTZII, Brongn. (*C. irregularis*, L. & H.)

M. C., Joggins (*J. W. D.*).

Often erect; has large irregular adventitious roots. This species is regarded by Brongniart as probably belonging to *Calamodendron*.

6. C. DUBIUS, Artis.

M. C., Sydney (*R. Brown*); Joggins (*J. W.D.; Logan*); U. C., Pictou (*J. W. D.*).

7. C. NOVA-SCOTICA, spec. nov.   Pl. XII. fig. 89.

M. C., Joggins (*J. W. D.*).

Ribs equal, less than a line wide, striated longitudinally. Joints obscurely marked, and with circular areoles separated by the breadth of three to four ribs. Bark of moderate thickness.

8. C. NODOSUS, Schloth.

This species has long slender branchlets, with close whorls of short rigid leaves.

M. C., Sydney (*R. Brown*); Grand Lake (*C. F. Hartt*).

9. C. ARENACEUS (?), Jäger.

This species is mentioned with doubt in Lyell's list.

EQUISETITES, Sternberg.

EQUISETITES CURTA, spec. nov.   Pl. XII. fig. 88.

Short thick stems, enlarging upward, and truncate above; joints numerous; sheaths as long as the joints, with unequal acuminate keeled points. Lateral branches or fruit with longer leaf-like points. Has the characters of *Equisetites*; but its affinities are quite uncertain.

M. C., Sydney (*R. Brown*).

M 2

ASTEROPHYLLITES, Brongn.

1. ASTEROPHYLLITES FOLIOSA, L. & H.

M. C., Joggins (*J. W. D.*); Sydney (*R. Brown*).

2. A. EQUISETIFORMIS, L. & H.

M. C., Sydney (*R. Brown*); Pictou (*J. W. D.*).

3. A. GRANDIS, Sternberg?

The specimens resemble this species, but are not certainly the same. Logan's specimens have terminal spikes of fructification.

M. C., Grand Lake (*C. F. Hartt*); Bay de Chaleur (*Logan*); Sydney (*Bunbury*).

4. ASTEROPHYLLITES, sp.

A species with tubercles (fruit) in the axils is mentioned in Lyell's list as from Sydney. I have not seen it, but have a specimen from Mr. Brown similar to *A. tuberculata*, Sternberg, which may be the same.

5. A. TRINERVIS, spec. nov. Pl. XIII. fig. 90.

Main stem smooth, delicately striate, with leaves at the nodes. Branches delicately striate, with numerous whorls of linear nearly straight leaves, 0·5 inch long, twenty or more in a whorl, and showing two lateral nerves in addition to the median nerve. This and *A. equisetiformis* would be placed by some authors in *Annularia*.

M. C., Sydney (*R. Brown*).

ANNULARIA, Sternberg.

ANNULARIA GALIOIDES, Zenker.

M. C., Grand Lake (*C. F. Hartt*); U. C., Pictou (*J. W. D.*); Bay de Chaleur (*Logan*); Sydney (*R. Brown*).

SPHENOPHYLLUM, Brongn.

1. SPHENOPHYLLUM EMARGINATUM, Brongn.

M. C., Sydney (*R. Brown*); Grand Lake (*C. F. Hartt*); Bay de Chaleur (*Logan*); Pictou (*J. W. D.*).

2. S. LONGIFOLIUM, Germar.

U. C., Pictou (*J. W. D.*); M. C., Sydney (*R. Brown*).

3. S. SAXIFRAGIFOLIUM, Sternberg.

Elongate much-forked variety, closely allied to *S. bifurcatum*, Lesquereux.

Bay de Chaleur (*Logan*).

4. S. SCHLOTHEIMII, Brongn.

M. C., Sydney (*Bunbury*).

5. S. EROSUM, L. and H.

M. C., Sydney (*Bunbury*).

The last two species are regarded by Geinitz as varieties of *S. emarginatum*. A specimen of the last-named species in Sir William

Logan's collection shows a woody jointed stem like that of *Astero-phyllites*, giving off branches at the joints; these again branch and bear whorls of leaves. The stem shows under the microscope a single bundle of reticulated or scalariform vessels like those of some ferns, and also like those of *Tmesipteris*, as figured by Brongniart. This settles the affinities of these plants as being with ferns or with *Lycopodiaceæ*.

PINNULARIA, L. & H.

1. PINNULARIA CAPILLACEA, L. & H.

M. C., Sydney (*R. Brown*).

2. P. RAMOSISSIMA, spec. nov.

More slender and ramose than the last.

M. C., Joggins (*J. W. D.*).

3. P. CRASSA, spec. nov.

Branching like *P. capillacea*, but much stronger and coarser.

L. C., Horton (*C. F. Hartt*).

All these are apparently branching fibrous roots, of soft cellular tissue with a thin epidermis and slender vascular axis. Perhaps they are roots of *Asterophyllites*.

Genus NŒGGERATHIA, Steinberg.

1. NŒGGERATHIA DISPAR, spec. nov.   Pl. XIII. fig. 91.

A remarkable fragment of a leaf, with a petiole nearly 3 inches long, and a fourth of an inch wide, spreading abruptly into a lamina, one side of which is much broader than the other, and with parallel veins running up directly from the margin as from a marginal rib. It appears to be doubled in at both edges, and is abruptly broken off. It seems to be a new species; but of what affinities, it is impossible to decide.

Bay de Chaleur (*Sir W. E. Logan*).

2. N. FLABELLATA, L. & H.

M. C., Sydney (*R. Brown*).

CYCLOPTERIS, Brongn.

(including *Cyclopteris* proper, and subgenera *Aneimites*, Daws., and *Neuropteris*, Brongn.).

1. CYCLOPTERIS HETEROPHYLLA, Goeppert.

M. C. and U. C., Joggins (*J. W. D.*).

2. C. (ANEIMITES) ACADICA, Dawson, Quart. Journ. Geol. Soc. vol. xvii. p. 5. Pl. VIII. fig. 32.

Stipe large, striate, branching dichotomously several times. Pinnæ with several broadly obovate pinnules grouped at the end of a slender petiolule, and with dichotomous radiating veins. Fertile pinnæ with recurved petiolules, and borne on the divisions of the main petiole near their origin. This plant might be placed in the genus *Adiantites*, Brongn., but for the fructification, which allies it with such forms as *Aneimia*. It has a very large frond, the main petiole being sometimes 3 inches in diameter, and 2 feet long before branching. Flattened petioles have sometimes been mistaken for *Cordaites* and *Schizopteris*. It is a characteristic plant of the Lower Coal-measures.

L. C., Horton (*C. F. Hartt*); Norton Creek, N.B. (*G. F. Matthew*).

3. C. OBLONGIFOLIA, Goeppert.

A little larger and coarser than Goeppert's figure.

U. C., Pictou (*J. W. D.*).

4. C. (NEUROPTERIS) OBLIQUA, Brongn.

M. C., Sydney (*R. Brown*); Grand Lake (*C. F. Hartt*).

5. C. (?NEUROPTERIS) INGENS, L. & H.

M. C., Sydney (*R. Brown*); Grand Lake (*C. F. Hartt*).

6. C. OBLATA, L. & H.

M. C., Sydney (*R. Brown*).

7. C. FIMBRIATA, Lesquereux.

M. C., Sydney (*R. Brown*).

8. C. HISPIDA, spec. nov.   Pl. XIII. fig. 92.

Pinnate; pinnules obovate, diminishing in size towards the point, decurrent on the petiole; veins slender, distant, forking several times; under surface covered with stiff hairs.

M. C., Sydney (*R. Brown*).

9. C. ANTIQUA, spec. nov.   Pl. XIII. fig. 95.

L. C., ? Hebert River (*J. W. D.*).

Tripinnate; petioles slender; pinnules oblong, obtuse, decurrent on the petiole, not contiguous. Terminal pinnules much elongated; venation simple, divergent. This plant approaches more nearly to the peculiar species of *Cyclopteris* found in the Devonian, than any of the others I have seen in the Carboniferous.

NEUROPTERIS, Brongn.

1. NEUROPTERIS RARINERVIS, Bunbury.

M. C., Sydney (*R. Brown*); Grand Lake (*C. F. Hartt*); Bay de Chaleur (*Logan*).

2. N. PERELEGANS, spec. nov.   Pl. XIII. fig. 93.

M. C., Sydney (*R. Brown*).

Resembles *N. elegans*, Brongn., but has narrower pinnules, and nerves less oblique to the midrib. The pinnules were thick and leathery, rough or cellular-netted above, and showing the venation only on the underside.

3. N. CORDATA, Brongn. (and var. *angustifolia*).

The ferns referred to this species are identical with *N. hirsuta* of Lesquereux. They abound in the Middle and Upper Coal-formations, and have larger pinnules than any of the other ferns.   A

single terminal pinnule in my collection is 5 inches long. The surface is always more or less hairy.

M. C., Sydney (*R. Brown*) ; U. C., Pictou (*J. W. D.*).

4. N. Voltzii, Brongn.

A single imperfect specimen like this species, but uncertain.
M. C., Pictou (*J. W. D.*).

5. N. gigantea, Stornb.

M. C., Sydney (*R. Brown*) ; Grand Lake (*C. F. Hartt*); U. C., Pictou (*J. W. D.*).

6. N. flexuosa, Stornb.

M. C., Sydney (*R. Brown*) ; Joggins (*J. W. D.*).

7. N. heterophylla, Brongn.

M. C., Sydney (*R. Brown*); U. C., Pictou (*J. W. D.*).

9. N. Loshii, Brongn.

Bay de Chaleur (*Logan*).

10. N. acutifolia, Brongn.

M. C., Sydney (*Lyell's list*).

11. N. conjugata, Goepp.

M. C., Sydney (*Brown's list, Acad. Geol.*).

12. N. attenuata, L. & H.

M. C., Sydney (*l. c.*).

13. N. dentata, Lesq.

M. C., Sydney (*R. Brown*).

14. N. Soretii (Brongn.).

M. C., Sydney (*R. Brown*).

15. N. auriculata, Brongn.

M. C., Sydney (*R. Brown*).

16. N. cyclopteroides, spec. nov.   Pl. XIII. fig. 94.

Pinnate; pinnules contiguous or overlapping, obliquely round-
ovate, attached at the lower third of the base; nerves numerous,
spreading from the point of attachment. Allied to *N. Villiersi*,
Brongn.

M. C., Sydney (*R. Brown*).

Odontopteris, Brongn.

1. Odontopteris Schlotheimii, Brongn.

M. C., Sydney (*R. Brown*) ; Bay de Chaleur (*Logan*); U. C., Pictou
(*J. W. D.*).

2. O. subcuneata, Bunbury.

M. C., Sydney (*R. Brown*).

156 PROCEEDINGS OF THE GEOLOGICAL SOCIETY. [Dec. 20,

DICTYOPTERIS, Gutb.

DICTYOPTERIS OBLIQUA, Bunbury.
M. C., Sydney (*R. Brown*).

LONCHOPTERIS, Brongn.

LONCHOPTERIS TENUIS, spec. nov. Pl. XIII. fig. 103.
Pinnate or bipinnate; pinnules contiguous at the base, nearly at
right angles to petiole, oblong elongate, obtuse. Network of veins
very delicate. Allied to *L. Bricii*, Brongn., but with smaller, more
elongate pinnules and finer veins. I suspect this to be a thick-leaved
Pecopteris, showing a coarse cellular reticulation on the upper surface.
M. C., Sydney (*R. Brown*).

SPHENOPTERIS, Brongn.

1. SPHENOPTERIS MUNDA, spec. nov. Pl. XIII. fig. 97.
Like *S. Dubuissonii*, Brongn., or *S. irregularis*, Sternberg, in habit;
but the pinnules are obovate, decurrent, and few-veined.
M. C., Grand Lake (*C. F. Hartt*).

2. S. HYMENOPHYLLOIDES, Brongn.
M. C., Sydney (*R. Brown*); U. C., Joggins (*J. W. D.*).

3. S. LATIOR, spec. nov. Pl. XIII. fig. 98.
Petiole forking at an obtuse angle, slender, tortuous; divisions
bipinnate; pinnæ with broad, rounded, confluent pinnules; veins
twice forked, with sori in the forks of the veins. In habit like *S.
latifolia*, Brongn., *S. Newberryi*, and *S. squamosa*, Lesq.
M. C., Grand Lake (*C. F. Hartt*); U. C., Pictou (*J. W. D.*).

4. S. DECIPIENS, Lesquereux.
M. C., Sydney (*R. Brown*).

5. S. GRACILIS, Brongn.
M. C., Joggins (*J. W. D.*); Grand Lake (*C. F. Hartt*).

6. S. ARTEMISIÆFOLIA, Brongn.
M. C., Grand Lake (*C. F. Hartt*); Sydney (*R. Brown*).

7. S. CANADENSIS, spec. nov. Pl. XIII. fig. 99.
General aspect like *S. Hœninghausi*, but secondary pinnules with
a margined petiole, and oblong pinnules divided into three to five
obtuse points. It is not unlike *S. marginata*, from the Devonian of
St. John.
Bay de Chaleur (*Logan*); Sydney? (*R. Brown*).

8. S. LESQUEREUXII, Newberry.
M. C., Sydney (*R. Brown*).

9. S. MICROLOBA, Guttbier.
M. C., Sydney (*R. Brown*).

10. S. OBTUSILOBA (?), Brongn.
M. C., Bay de Chaleur (*Logan*).

PHYLLOPTERIS, Brongn.

PHYLLOPTERIS ANTIQUA, spec. nov.   Pl. XIII. fig. 98

Pinnate; petiole thick, woody; pinnules oblong, pointed, attached by the middle of the base; midrib strong, extending to the point, giving off very oblique nerves, which have obliquely pinnate nervules not anastomosing.   A remarkable frond, which, if not the type of a new genus, must belong to that above named.

M. C., Sydney (*R. Brown*).

ALETHOPTERIS, Sternberg.

1. ALETHOPTERIS LONCHITICA, Sternberg.

M. and U. C., Joggins (*J. W. D.*); M. C., Sydney (*R. Brown*); Grand Lake (*C. F. Hartt*).

Very abundant throughout the Middle and Upper Coal-formations, and so variable that several species might easily be founded on detached specimens.

2. A. HETEROPHYLLA, L. & H.

L. C., Parrsborough (*A. Gesner*).

3. A. GRANDINI, Brongn.

M. C., Sydney (*R. Brown*).

4. A. NERVOSA, Brongn.

M. C., Sydney (*R. Brown*); Bay de Chaleur (*Logan*); U. C., Pictou (*J. W. D.*).

5. A. MURICATA, Brongn.

M. C., Joggins, Bathurst (*Lyell*); U. C., Pictou (*J. W. D.*).

6. A. PTEROIDES, Brongn. (*A. Brongniartii*, Goeppert).

L. or M. C., Bathurst (*Lyell's list*).

7. A. SERLII, Brongn.

M. C., Sydney (*R. Brown*); Bay de Chaleur (*Logan*).

8. A. GRANDIS, spec. nov.   Pl. XIII. fig. 100.

Bipinnate; pinnæ broad, contiguous, united at the base; veins numerous, once forked, not quite at right angles to the midrib. Upper pinnæ having the pinnules confluent so as to give crenate edges.   Still higher the apex of the frond shows distant decurrent long pinnules with waved margins.   A very large and fine species of the type of *A. Serlii* and *A. Grandini*, but much larger and different in details.   Its texture seems to have been membranaceous; and fragments from that part of the frond where the long simple pinnules are passing into the compound ones might be mistaken for an *Odontopteris*.

Bay de Chaleur (*Logan*).

PECOPTERIS, Brongn.

1. PECOPTERIS ARBORESCENS, Schloth.

Seems to have been an herbaceous species with a very strong

petiole. It occurs in an erect position in a sandstone on Wallace River.

M. C., Sydney (*R. Brown*); U. C., Pictou (*J. W. D.*); Wallace River (*Dr. Creed*).

2. P. ABBREVIATA, Brongn.

M. C., Sydney (*R. Brown*); Salmon River, U. C., Pictou (*J. W. D.*). Very common both in the Upper and Middle Coal-formations.

3. P. RIGIDA, spec. nov.

Similar to *P. arborescens*, but much smaller, and with finer nerves. U. C., Pictou (*J. W. D.*).

4. P. UNITA, Brongn.

Certain pinnules of a frond are sometimes swollen as if covered with fructification below; and in this state they resemble *P. arguta*, Brongn. The sori are seen in other specimens, and are large, round, and covered with an indusium as in *Aspidium*.

M. C., Sydney (*R. Brown*); U. C., Pictou (*J. W. D.*).

5. P. PLUMOSA, Brongn.

M. C., Sydney (*R. Brown*).

6. P. POLYMORPHA, Brongn.

M. C., Sydney (*R. Brown*).

7. P. ACUTA, Brongn.

M. C., Pictou (*J. W. D.*).

8. P. LONGIFOLIA, Brongn.

In Bunbury's list, from Sydney.

9. P. TÆNIOPTEROIDES, Bunbury.

M. C., Sydney (*R. Brown*).

10. P. CYATHEA, Brongn.

M. C., Sydney (*R. Brown*).

11. P. ÆQUALIS, Brongn.

M. C., Sydney (*R. Brown*).

12. P. SILLIMANI?, Brongn.

In Lyell's list, from Sydney.

13. P. VILLOSA, Brongn.

M. C., Pictou (*Lyell's list*).

14. P. BUCKLANDI, Brongn.

M. C., Sydney (*Brown's list*).

15. P. OREOPTEROIDES, Brongn.

M. C., Sydney (*Brown's list*).

16. P. DECURRENS, Lesq.

Has pinnules more crowded, decreasing towards the apex, but may be a variety.

M. C., Sydney (*R. Brown*).

17. P. PLUCKENETII, Ster ᵇ.

M. C., Sydney (*R. Brown*).

BEINERTIA, Goeppert.

BEINERTIA GŒPPERTI, spec. nov.   Pl. XIII. fig. 101.

Bipinnate; pinnæ broad, contiguous, obtuse, with thick pinnules. Pinnules rounded above, obovate below. Midrib thick, oblique, dividing above into a tuft of irregular hair-like veins.

M. C., Grand Lake (*C. F. Hartt*); Bay de Chaleur (*Logan*); U. C., Joggins (*J. W. D.*).

HYMENOPHYLLITES, Goeppert.

HYMENOPHYLLITES PENTADACTYLA, spec. nov.

In general habit like *Sphenopteris microloba*, Goepp., but with pinnules divided into from four to seven obtuse cuneate lobes, each with one vein.

M. C., Sydney (*R. Brown*).

PALÆOPTERIS, Geinitz.

1. PALÆOPTERIS HARTII, spec. nov.   Pl. VIII. fig. 35.

Stem or leaf-bases transversely wrinkled with delicate lines; scars transversely oval, slightly appendaged below; vascular scars confluent. Breadth 1·4 in.; length 0·6 in.

M. C., Grand Lake (*C. F Hartt*).

2. P. ACADICA, spec. nov.   Pl. VIII. fig. 36.

Stem or leaf-bases longitudinally striated; scars transverse, flat above, rounded and bluntly appendaged below; vascular scars in a transverse row. Breadth of scars 0·7 inch; length 0·5 inch.

U. C., Pictou (*J. W. D.*).

CAULOPTERIS, L. & H.

Several small erect stems at the Joggins seem to be trunks of ferns, but are too obscure for description.

PSARONIUS, Cotta.

Trunks of this kind must be rare in the Nova Scotian coal-fields. A few obscure stems surrounded by cord-like aërial roots have been found, and probably are remains of plants of this genus.

MEGAPHYTON, Artis.

1. MEGAPHYTON MAGNIFICUM, spec. nov.   Pl. VIII. fig. 34.

Stems large, roughly striated longitudinally; scars contiguous, orbicular, deeply sunk, nearly 3 inches in diameter, and each with a bilobate vascular impression 2 inches broad and an inch high.

M. C., Joggins (*J. W. D.*).

2. M. HUMILE, spec. nov. Pl. VIII. fig. 33.

Stem 2·5 inches in diameter; leaf-scars prominent, flattened, and broken at the ends, 1 inch wide. Surface of the stem marked with irregular furrows, and invested with a carbonaceous coating. An internal axis, nearly 2 inches in diameter, with a coaly coating, sends off obliquely thick branches to the leaf-scars. This is a very remarkable specimen, and throws much light on the structure of *Megaphyton*. Unfortunately the minute structures are not preserved.

M. C., Sydney (*R. Brown*).

Genus LEPIDODENDRON, Sternberg.

1. LEPIDODENDRON CORRUGATUM, Dawson, Quart. Journ. Geol. Soc. vol. xv. Pl. XI. fig. 53.

Areoles elongate ovate, acute at both ends, with a ridge along the middle, terminating in a single elevated vascular scar at the upper end. In certain states the vascular mark appears in the middle of the areole. In young branches the areoles are contiguous and resemble those of *L. elegans*. In old stems they become separated by spaces of longitudinally wrinkled bark; in very old stems these spaces are much wider than the areoles. Leaves linear, 1 inch or more in length, usually reflected, one-nerved. Cones (*Lepidostrobi*) terminal, short, cylindric, with numerous short, acute-triangular scales. Structure of stem:—a central pith with a slender cylinder of scalariform vessels, exterior to which is a thick cylinder of cellular tissue and bast fibres, and a dense outer bark.

Var. *verticillatum* has the areoles arranged in regular decussate whorls instead of spirally. This difference, which might at first sight seem to warrant even a generic distinction, is proved by specimens in my possession to be merely a variety of *phyllotaxis*.

This species is eminently characteristic of the Lower Carboniferous Coal-measures, and has not yet been found in the Middle Coal-formation. Fragments of bark resembling that of this species, occur in the Coal-formation of Bay de Chalour, along with leafy branches of *Lepidodendron*, which resemble those of this species, though, I believe, distinct.

L. C., Horton, &c. (*C. F. Hartt; J. W. D.*); Norton Creek, &c., New Brunswick (*G. F. Matthew*).

2. L. PICTOENSE, spec. nov. Pl. IX. fig. 37.

Areoles contiguous, prominent, long oval, acuminate, separated in young stems by a narrow line; breadth to length as 1 to 3, or less; lower half obliquely wrinkled, especially at one side. Middle line indistinct. Leaf-scar at upper end of areole, small, triangular, with traces of three vascular points, nearly confluent. Length of areole about 0·5 inch. Leaves contracted at the base, widening slightly, and gradually contracting to a point; ribs three, central distinct, lateral obscure; length 1 inch. Cones borne at the extremities of the smaller branches, oblong, obscurely scaly.

In habit of growth this species resembles *L. elegans*, for which imperfect specimens might be mistaken.   It is also near to *L. binerve* and *L. patulum*, Bunbury*.   It abounds in the Middle Coal-measures.

M. C., Sydney (*R. Brown*); Pictou (*H.Poole and J. W.D.*); Grand Lake (*C. F. Hartt*).

3. L. RIMOSUM, Sternberg.

M. C., Sydney (*R. Brown*); Joggins (*J. W. D.*).

4. L. DICHOTOMUM, Sternberg (*L. Sternbergii*, L. & H.).

M. C., Sydney (*R. Brown*); Joggins (*J. W. D.*); L. C., Horton, (*J. W. D.*).

5. L. DECURTATUM, spec. nov.   Pl. IX. fig. 40.

Areoles approximate or separated by a shallow furrow, rhombic ovate, obliquely acuminate below, nearly as broad as long, wrinkled transversely, especially on the middle line, which appears tuberculated; vascular scar rhombic, twice as broad as long, with three approximate vascular points.   In some flattened specimens the line separating the areoles is indistinct, and the scars appear on a transversely wrinkled surface without distinct areoles.

M. C., Pictou (*J. W. D.*).

6. L. UNDULATUM, Sternberg.   Pl. IX. fig. 41.

Possibly several species are included under this name; but they cannot be separated at present.

M. C., Sydney (*R. Brown*); Joggins and Pictou (*J. W. D.*); U. C., Joggins (*J. W. D.*).

7. L. DILATATUM, Lindley & Hutton.

M. C., Joggins (*J. W. D.*).

8. L., sp. like TETRAGONUM, Goepp.

Obscurely marked, but a distinct species, unless an imperfectly preserved variety of *L. tetragonum.*   The areoles are square, with a rhombic scar at the upper corner of each.

L. C., Horton (*J. W. D.*).

9. L. BINERVE, Bunbury.

M. C., Sydney (*R. Brown*).

10. L. TUMIDUM, Bunbury.

I think it probable that this species belongs to the genus *Lepidophloios*; but I have not seen a specimen.

M. C., Sydney (*R. Brown*).

11. L. GRACILE, Brongn.

In Brown's list in ' Ac. Geology.'   Probably a variety of the next.

M. C., Sydney (*R. Brown*).

12. L. ELEGANS, Brongn.

In Bunbury and Brown's lists.

M. C., Sydney (*R. Brown*).

---

* In certain states of preservation, the lateral ribs of the leaves become obsolete; and in others the central disappears, in which state the resemblance to *L. binerve* is very close.

13. L. PLUMARIUM, L. & H.
   M. C., Sydney (*in Brown's list*).
14. L. SELAGINOIDES, Sternb.
   M. C., Sydney (*in Brown's list*).
15. L. HARCOURTII (Witham).
   M. C., Sydney (*in Brown's list*).
16. L. CLYPEATUM (?), Lesqx.
   M. C., Sydney (*R. Brown*); U. C., Joggins (*J. W. D.*).
17. L. ACULEATUM, Sternberg.
   M. C., Sydney (*R. Brown*).
18. L. PLICATUM, spec. nov. Pl. IX. fig. 38.
   Leaf-areoles much elongated; breadth to length as 1 to 5 or 6, transversely rugose; central line indistinct. Leaf-scar rhombic, with three vascular points; scars in old stems separated by rugose bark, and somewhat elongate.
   M. C., Pictou (*J. W. D.*).
19. L. PERSONATUM, spec. nov. Pl. IX. fig. 39.
   Areoles ovate acuminate; breadth to length as 1 to 3 or 4, contiguous in young stems; central lines distinct; lower part of areole with transverse lines. Leaf-scars ovate, with two marks above and two below; leaves slender, 1 inch long, one-nerved.
   M. C., Sydney (*R. Brown*).

HALONIA, sp.   HALONIA, L. & H.
   A specimen probably referable to this genus from Grand Lake, in the collection of C. F. Hartt.

LEPIDOSTROBUS, Brongn.

1. LEPIDOSTROBUS VARIABILIS, L. & H.
   The most common species.
   M. C., Sydney (*R. Brown*); Pictou and Joggins (*J. W. D.*).
2. L. SQUAMOSUS, spec. nov. Pl. 10. fig. 46.
   2 to 3 inches long, 1 inch thick; scales large, broadly trigonal, acute. Allied to *L. trigonolepis*, but larger. Probably a cone of *Lepidophloios*.
   M. C., Grand Lake (*C. F. Hartt*).
3. L. LONGIFOLIUS, spec. nov.
   Long-leaved, like *Lepidodendron longifolium*, L. & H.
   M. C., Joggins (*J. W. D.*).
4. LEPIDOSTROBUS, sp.
   Acute trigonal leaves, small.
   M. C., Joggins (*J. W. D.*).
5. LEPIDOSTROBUS, sp.
   Round, with obscure scales and remains of long leaves.
   L. C., Horton (*J. W. D.*).

6. L. TRIGONOLEPIS, Bunbury.
   M. C., Sydney (*R. Brown*).

LEPIDOPHYLLUM, Brongn.

1. LEPIDOPHYLLUM LANCEOLATUM, L. & H.
   M. C., Joggins; U. C., Pictou (*J. W. D.*).

2. L. TRINERVE (?), L. & H.
   Two-nerved or three-nerved, like *L. trinerve*, L. & H., but narrower. Both the above are parts of *Lepidostrobi.*
   U. C., Joggins (*J. W. D.*).

3. L. MAJUS (?), Brongn.
   M. C., Sydney (*R. Brown*).

4. LEPIDOPHYLLUM, sp.
   Broad ovate, short, pointed, one-nerved, half an inch long.
   U. C., Pictou.

5. L. INTERMEDIUM, L. & H.
   M. C., Sydney (*R. Brown's list*).
   Halonia, Lepidostrobus and *Lepidophyllum*, including only parts of *Lepidodendron* and *Lepidophloios*, are to be regarded as merely provisional genera.

LEPIDOPHLOIOS, Sternberg.

1. LEPIDOPHLOIOS ACADIANUS, spec. nov. Pl. X. fig. 45, Pl. XI.
   fig. 51.
   Leaf-bases broadly rhombic, or in old stems regularly rhombic, prominent, ascending, terminated by very broad rhombic scars having a central point and two lateral obscure points. Outer bark laminated or scaly. Surface of inner bark with single points or depressions. Leaves long, linear, with a strong keel on one side, 5 inches or more in length. Cone-scars sparsely scattered on thick branches, either in two rows or spirally, both modes being sometimes seen on the same branch. Scalariform axis scarcely an inch in diameter in a stem 5 inches thick. Fruit, an ovate strobile with numerous acute scales covering small globular spore-cases. This species is closely allied to *Ulodendron majus* and *Lepidophloios laricinus*, and presents numerous varieties of marking.
   M. C., Joggins, Salmon River, Pictou (*J. W. D.*); Sydney (*R. Brown*).

2. L. PROMINULUS, spec. nov. Pl. XI. fig. 52.
   Leaf-bases rhombic pyramidal, somewhat wrinkled at the sides, truncated by regularly rhombic scars, each with three approximate vascular points.
   M. C., Joggins (*J. W. D.*).

3. L. PARVUS, spec. nov. Pl. XI. fig. 50.
   Leaf-bases rhombic, small, with rhombic scars broader than long;

vascular points obscure; leaves linear, acute, 3 inches or more in length, with a keel and two faint lateral ribs. Cones large, scssile. U. C., Pictou; M. C., Joggins (*J. W. D.*); M. C., Sydney (*R. Brown*).

4. L. PLATYSTIGMA, spec. nov. Pl. X. figs. 47 & 48.

Leaf-bases rhombic, broader than long, little prominent; scars rhombic, oval, acuminate, slightly emarginate above; vascular points two, approximate or confluent.
M. C., Sydney (*R. Brown*); Joggins (*J. W. D.*).

5. L. TETRAGONUS, spec. nov. Pl. X. fig. 49.

Leaf-bases square, furrowed on the sides; leaf-scar central, with apparently a single central vascular point.
M. C., Joggins (*J. W. D.*).

DIPLOTEGIUM, Corda.

DIPLOTEGIUM RETUSUM, spec. nov. Pl. XIII. fig. 102.

The fragments referable to plants of this genus are imperfect and obscure. The most distinct show leaf-bases ascending obliquely, and terminating by a retuse end with a papilla in the notch. Some less distinct fragments may possibly be imperfectly preserved specimens of *Lepidodendron* or *Lepidophloios*.
M. C., Joggins (*J. W. D.*).

KNORRIA.

Nearly all the plants referred to this genus in the Carboniferous rocks are, as Goeppert has shown, imperfectly preserved stems of *Lepidodendron*. In the Lower Coal-formation many such *Knorria* forms are afforded by *L. corrugatum*.

KNORRIA SELLONII, Sternberg.

This appears different from the ordinary *Knorrice*; its supposed leaves may be aërial roots. It has a large pith-cylinder with very distant tabular floors, like *Sternbergia*.
M. C., Sydney (*R. Brown*).

CORDAITES, Unger. (PYCHNOPHYLLUM, Brongn.)

1. CORDAITES BORASSIFOLIA, Corda.

M. C., Pictou (*H. Poole*); Grand Lake (*C. F. Hartt*); Sydney (*R. Brown*); Joggins, Onslow (*J. W. D.*); Bay de Chaleur (*Logan*).
Very abundant in the Middle Coal-formation.

2. C. SIMPLEX, spec. nov.

Leaves similar to the last in size and form, but with simple, equal, parallel nerves. It may be a variety, but is characteristic of the Upper Coal-formation.
M. C., Grand River (*C. F. Hartt*); U. C., Pictou (*J. W. D.*).

CARDIOCARPUM, Brongn.

1. CARDIOCARPUM FLUITANS, spec. nov. Pl. XII. fig. 74.

Oval; apex entire or notched; surface slightly rugose; nucleus round ovate, acuminate, pitted on the surface, with a raised mesial line. M. C., Joggins (*J. W. D.*).

2. C. DISLCTUM, spec. nov. Pl. XII. fig. 73.

Nucleus as in the last species, but striate; margin widely notched at apex, and more narrowly notched below. M. C., Grand Lake (*C. F. Hartt*).

3. CARDIOCARPUM, sp. like *C. marginatum*. M. C., Joggins (*J. W. D.*).

4. CARDIOCARPUM, sp. allied to *C. latum*, Newberry. M. C., Pictou (*H. Poole*).

These *Cardiocarpa* are excessively abundant in the roofs of some coal-seams; and the typical ones must have been samaras or winged nutlets. They must have belonged to phœnogamous plants, and certainly are not the fruits of *Lepidodendron*, though some of the spore-cases of this genus have been described as *Cardiocarpa*. These I propose to place under the provisional genus *Sporangites*.

SPORANGITES, Dawson.

1. SPORANGITES PAPILLATA, spec. nov. Pl. XII. fig. 80.

I propose the provisional generic name of *Sporangites* for spores or spore-cases of *Lepidodendron*, *Calamites*, and similar plants, not referred to the species to which they belong. The present species is round, about 1 inch in diameter, and covered with minute raised papillæ or spines. It abounds in the roof of several of the shaly coals in the Joggins section, and especially in one in group 19 of that section. M. C., Joggins (*J. W. D.*).

2. S. GLABRA, spec. nov. Pl. XII. fig. 81.

About the size of a mustard-seed, round and smooth. Exceedingly abundant in the Lower Carboniferous Coal-measures of Horton Bluff, with *Lepidodendron corrugatum*, to which it possibly belongs. A similar spore-case, possibly of another species of *Lepidodendron*, occurs rarely in the Middle Coal-formation at the Joggins.

STERNBERGIA, Artis.

This provisional genus includes the piths of *Dadoxylon*, *Sigillaria*, and other plants, usually preserved as casts in sandstone, retaining more or less perfectly the transverse partitions into which the pith-cylinders of many coal-formation trees became divided in the process of growth. These fossils are most abundant in the Upper Coal-formation, but occur also in the Middle Coal-formation. The following varieties may be distinguished:—

(*a*) Var. *approximata*, with fine uniform transverse wrinkles. This is usually invested with a thin coating of structureless coal.

(*b*) Var. *angularis*, with coarser and more angular transverse wrinkles. This is the character of the pith of *Dadoxylon*.

(*c*) Var. *distans*, usually of small size, and with distant and irregular wrinkles. This is sometimes invested with wood having the structure of *Calamodendron*, and perhaps is not generically distinct from *C. approximatum*.

(*d*) Var. *obscura*, with distinct and distant transverse wrinkles, but not strongly marked on the surface. This is the character of the pith-cylinders of *Sigillaria* and *Lepidophloios*.

ENDOGENITES, L. & H.

Many sandstone-casts, answering to the character of the plants described under this name by Lindley, occur in the Upper Coal-formation. They are sometimes 3 inches in diameter, and several feet in length, irregularly striated longitudinally, and invested with coaly matter. Sometimes they show transverse striation in parts of their length. I believe they are casts of pith-cylinders of the nature of *Sternbergia*, and probably of Sigillarioid trees.

SOLENITES, L. & H.

Plants of this kind are found in the sandstones of the Upper Coal-formation of the Joggins.

For all the specimens noticed in the above list as collected by Sir W. E. Logan, Richard Brown, Esq., of Sydney, Cape Breton, Henry Poole, Esq., of Glace Bay, C.B., and G. F. and C. B. Matthew and C. F. Hartt, Esqs., St. John, New Brunswick, I am indebted to the kindness of those gentlemen. To Mr. Brown especially I am under great obligations for his liberality in placing at my disposal his large and valuable collection of the plants of the Cape Breton coal-field.

EXPLANATION OF PLATES V.-XIII.

Illustrative of the Coal-plants of British North America.

PLATE V.

Fig. 1. *Dadoxylon antiquius*: longitudinal section, radial, magnified 90 diameters.
2. The same: longitudinal section, tangential, magnified 90 diameters: *a*, medullary rays.
3. The same: portions of cells showing areolation, magnified 250 diameters.
4. *Dadoxylon Acadianum*: longitudinal section, radial, magnified 90 diameters.
5. The same: longitudinal section, tangential, magnified 90 diameters: *a*, medullary rays.
6. The same: portion of cell showing areolations, magnified 250 diameters.
7. *Dadoxylon materiarium*: longitudinal section, radial, magnified 70 diameters.
8. The same: longitudinal section, tangential, magnified 70 diameters: *a*, medullary rays.
9. The same: portion of shell showing areolation, magnified 250 diameters.
10. *D. annulatum*: longitudinal section, radial, magnified 70 diameters.

Fig. 11. The same: longitudinal section, magnified 90 diameters: *a*, one of the concentric rings of compact coaly matter.

12. The same: portion of a cell showing areolation, natural size.

13. The same: transverse section, natural size: *a*, pith; *b*, wood, composed of alternate circles of areolated cells and compact coaly matter; *c*, coaly bark.

PLATE VI.

Fig. 14. *Araucarites gracilis*: branch with leaves, three-fourths the natural size.

15. *Sigillaria Brownii*: portion of two ribs, 4 feet above the level, one-half the natural size: *a*, cortical; *b*, decorticated.

16. The same: portion of one rib near the root, furrowed and with vascular scars widely separated.

17. The same: trunk 1 foot in diameter at the base, showing the origin of the ribs.

18. The same: photograph of a portion of the upper part of the stem, one-half the natural size.

19. The same: portion of the base of the stem, one-half the natural size, from a photograph.

20. The same: Sigillarian root of this or an allied species.

21. *Sigillaria planicosta*, from a photograph, one-half the natural size; 21 *a*, scar and part of rib, one-half the natural size.

22. —— *catenoides*, from a photograph, one-half the natural size; 22 *a*, scar and part of ribs, one-half the natural size.

23. —— *striata*: flattened stem, from a photograph, two-thirds the natural size; 23 *a*, part of stem, same size; 23 *b*, scar, enlarged.

24. —— *eminens*: decorticated stem, from a photograph, much reduced; 24 *a*, scar, one-half the natural size.

25. —— *scutellata*, from a photograph, two-thirds the natural size, with band of interrupted growth; 25 *a*, ordinary areole, natural size; 25 *b*, areoles from the regular band, natural size.

PLATE VII.

Fig. 26. *Sigillaria elegans*: decorticated stem, from a photograph, one-half the natural size, with band of interrupted growth at β; 26 *a*, corticated and flattened stem, one-half the natural size, with branches at β & γ; 26 *b*, leaf of *S. elegans*; 26 *c*, areoles of a branch; 26 *d*, areole of main stem.

27. —— *Bretonensis*: corticated stem, from a photograph, two-thirds the natural size; 27 *a*, areole, natural size.

28. —— *Sydnensis*: decorticated stem, from a photograph, two-thirds the natural size; 28 *a*, areoles of decorticated stem; 28 *b* & *c*, variety of Stigmarian root attached to this species.

29. *Antholithes squamosus*, from a photograph, two-thirds the natural size.

30. —— *Rhabdocarpi*, from a photograph, two-thirds the natural size; 30 *a*, a larger specimen of the same; 30 *b*, single nutlet of the same, natural size.

30 *c*. —— *pygmæus*: fragment, natural size.

31. *Calamodendron approximatum*, one-half the natural size; 31 *a*, cast of pith; 31 *b*, coaly or woody investment; 31 *c*, tissues of wood of *Calamodendron*, magnified.

31 *d*. Tissues of a different species of *Calamodendron* or of *Calamites*.

PLATE VIII.

Fig. 32. Leaflets of *Cyclopteris Acadica*, from a photograph, one-third the natural size: 32 *a*, petiole of the same, one-third natural size; 32 *b*, divisions of petiole, one-half the natural size; 32 *c*, leaflets; 32 *d*, leaflet; 32 *e*, leaflet enlarged, showing venation; 32 *f*, striation of petiole enlarged; 32 *g*, *h*, remains of fructification; 32 *i*, base of petiole, much reduced.

33. *Megaphyton humile*: β, leaf-scars; γ, part of axis: from a photograph, two-thirds the natural size.

Fig. 34. *Megaphyton magnificum*: portion of stem, one-sixth the natural size;
          34 *a*, leaf-scar, from a photograph, two-thirds the natural size.
     35. *Palæopteris Hartii*, one-half the natural size.
     36. —— *Acadica*, one-half the natural size.

Plate IX.

Fig. 37. *Lepidodendron Pictoense*, branchlets and leaves, from a photograph,
          one-half the natural size (β, cone); 37 *a*, leafy branch; 37 *b*, branch
          with areoles; 37 *c*, larger branch; 37 *d*, old stem with deep furrows,
          and at β areoles, not enlarged; 37 *e*, leaves. Figs 37 *a* to 37 *e*, from
          photographs two-thirds the natural size. 37 *f*, areole enlarged; 37 *g*,
          leaf enlarged.
     38. —— *plicatum*, portion of old stem; 38 *a*, areole of branch, these areoles
          being placed in contact on such young branches.
     39. —— *personatum*, leafy branch; 39 *a*, larger stem with areoles; both
          from photographs, two-thirds the natural size; 39 *b*, areole, enlarged;
          39 *c*, leaf, natural size.
     40. —— *decurtatum*, from a photograph, two-thirds the natural size; 40 *a*,
          areole, enlarged.
     41. —— *undulatum*, portion of old stem, showing enlarged areoles, fur-
          rows, and two cone-scars.
     42, 43. Portions of old stems, probably of *L. rimosum* or allied species.

Plate X.

Fig. 44. Portions of old stems, probably of *Lepidodendron rimosum* or an allied
          species.
     45. *Lepidophloios Acadianus*, stem with marks of cones, from a photograph,
          one-half the natural size; 45 *a*, portion of stem with areoles, from
          a photograph, two-thirds the natural size; 45 *b*, decorticated stem,
          natural size; 45 *c* & *d*, opposite ⋅ of the same stem, reduced, to
          show the different arrangement of the cone-scars; 45 *e*, part of a
          leaf, natural size; 45 *f*, *g*, *h*, areoles from different parts of stem.
     46. Strobile of *Lepidophloios*; 46 *a*, transverse section of a similar strobile;
          from photographs, two-thirds the natural size.
     47 & 48. *Lepidophloios platystigma*, from a photograph, two-thirds the natural
          size; 47 *a* & 48 *a*, areole of the same, natural size.
     49. —— *tetragonus*, from photograph, two-thirds the natural size; 49 *a*,
          areole, two-thirds the natural size.

Plate XI.

Fig. 50. *Lepidophloios parvus*, stem with areoles and scars of cones; 50 *a*,
          group of leaves; both from photographs, two-thirds the natural size;
          50 *b*, areole, natural size.
     51. Cross section of *Lepidophloios Acadianus*, showing the outer rind and
          woody axis, one-tenth the natural size; 51 *a*, scalariform vessels of axis,
          magnified; 51 *b*, transverse section of part of the axis, showing the
          vascular bundles which proceed to the leaves, and the different
          diameters of the outer and inner circles of vessels; 51 *c*, smaller por-
          tion of the axis, showing one bundle of vessels.
     52. *Lepidophloios prominulus*, portion of cast, from a photograph, two-
          thirds the natural size; 52 *a*, areole, natural size.
     53. *Lepidodendron corrugatum*, young branch with cone; 53 *a*, branch with
          leaves; 53 *c*, older branch with areoles beginning to separate; 53 *d*,
          variety with alternate areoles; 53 *e*, variety with areoles in vertical
          rows; 53 *f*, *g*, old trunks, with widely separated areoles; 53 *h*, pho-
          tograph of branch; 53 *i*, Knorria, or decorticated state; 53 *k*, frag-
          ment, showing ramification; 53 *l*, bark with areoles in transverse rows;
          53 *m*, spore-case, natural size and magnified; 53 *n* to *r*, areoles in
          various states; 53 *s*, leaf, enlarged.
     54. Scalariform vessel of *Lepidodendron*.
     55.   „   „   *Stigmaria*.

PLATE XII.

Fig. 56. Scalariform vessel of *Lepidophloios*.
    57. Tissues of *Sigillaria*.
    58. Vessel of *Sphenophyllum*.
    59. Tissues of *Calamodendron*.
    60. Tissues of *Calamites*.
    61. Scalariform tissue of Ferns.
    62. Bast tissue of *Sigillaria*.
    63. Cells of *Dadoxylon Acadianum*.
    64. Cells of *D. materiarium*.
    65. Cuticle of *Pinnularia*.
    66. Vessel of *Sphenophyllum*, 200 diameters.
    67. Vessels and cells from vascular bundles of Ferns, 200 diameters.
    68. Tissues of *Sigillaria*, 200 diameters.
    69. *Rhabdocarpus insignis*.
    70–72. *Cardiocarpum*, spp.
    73. *C. bisectum*.
    74. *C. fluitans*.
    75. *Trigonocarpum minus*.
    76. *T. Sigillariæ*.
    77. *T. avellanum*.
    78. *T. intermedium* ; 78 *a*, nucleus of do.
    79. *T. Nœggerathi*.
    80. *Sporangites papillata* ; 80 *b*, nat. size.
    81. *S. glabra* ; 81 *b*, nat. size.
    82. Fragment of *Antholithes*.
    83. Stigmaria with scars in rhombic areoles.
    84. Stigmaria with bark divided by vertical furrows (var. *alternans*).
    85. Stigmaria with large scars in elongated areoles.
    86. Stigmaria with elongated scars (*Knorria* form).
    87. Stigmaria another variety, resembling *Diplotegium*.
    88. *Equisetites curtus*.
    89. *Calamites Nova-scoticus*.

Figs. 54–64 inclusive are drawn to a uniform scale of 90 diameters. Figs. 83 to 87 are taken from photographs, and are two-thirds the natural size.

PLATE XIII.

Fig. 90. *Asterophyllites trinervis* ; 90 *a*, portion of leaf enlarged.
    91. *Nœggerathia dispar*, one-half natural size.
    92. *Cyclopteris hispida* ; 92 *a*, portion of pinnule magnified, showing hairy surface and impressions of nervures.
    93. *Neuropteris perelegans* ; *a*, portion magnified, showing venation.
    94. —— *cyclopteroides*.
    95. *Cyclopteris antiqua*.
    96. *Phyllopteris antiqua* ; 96 *a*, portion magnified, showing venation.
    97. *Sphenopteris munda* ; 97 *a*, portion magnified.
    98. —— *latior* ; 98 *a*, pinnule magnified, showing venation and sori.
    99. —— *Canadensis* ; 99 *a*, pinnule magnified, showing venation.
    100 and 100 *a* & *b*. Pinnules of *Alethopteris grandis* ; 100 *a*, portion magnified, showing venation.
    101. *Beinertia Goepperti* ; 101 *a*, pinnule magnified, showing venation.
    102. *Diplotegium retusum* ; 102 *a*, leaf-scar magnified.
    103. *Lonchopteris tenuis* ; 103 *a*, portion enlarged, showing character of surface.

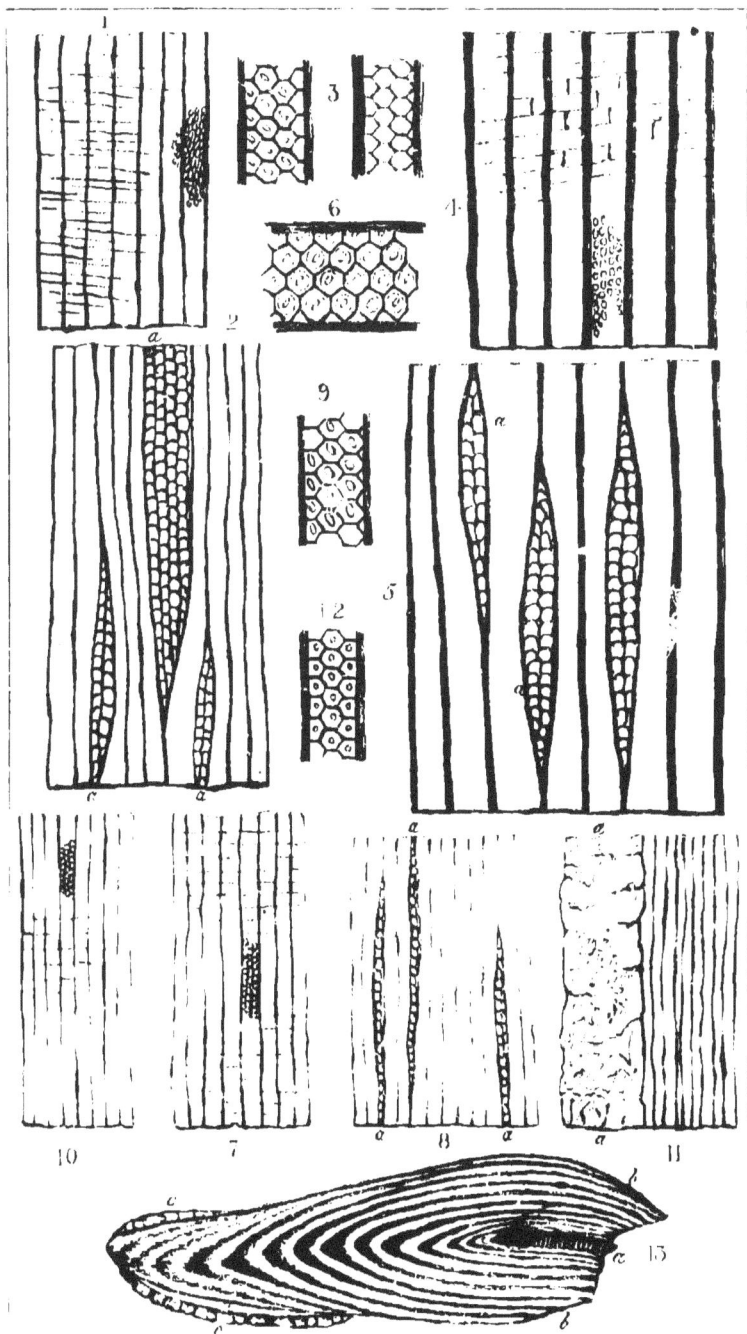

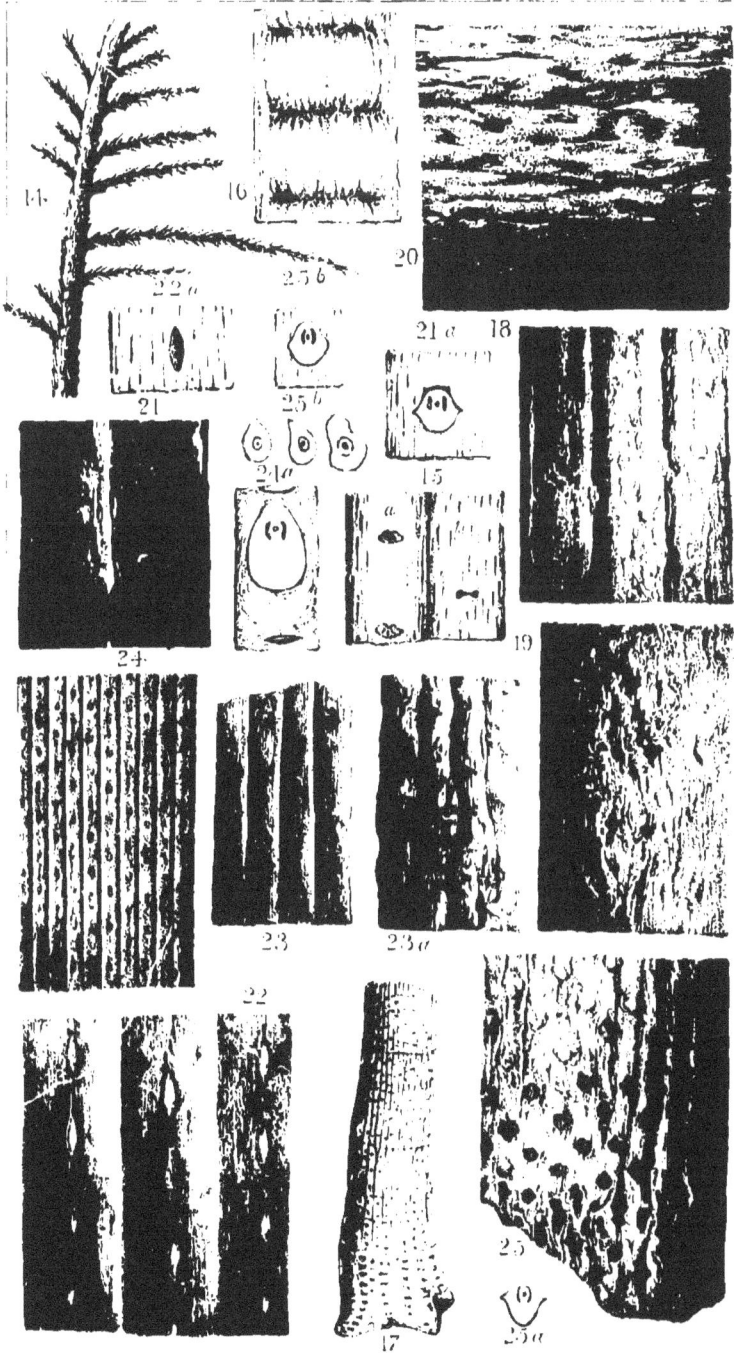

11.   16.   256   20.

22a   21a   18.

21.   25b   15.

21a

24.   19.

23.   23a

22.   25.

17.   25a

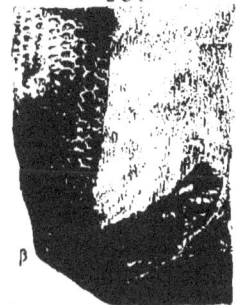

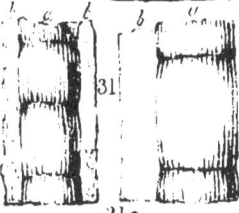

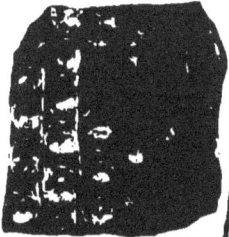

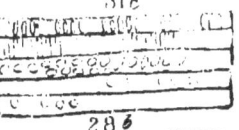

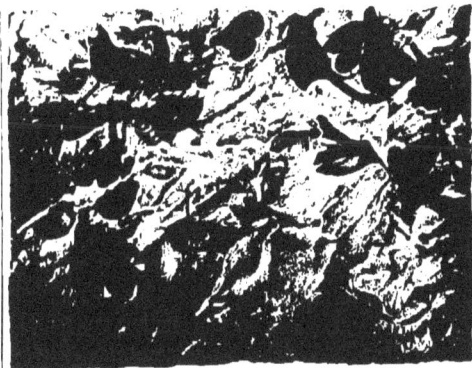

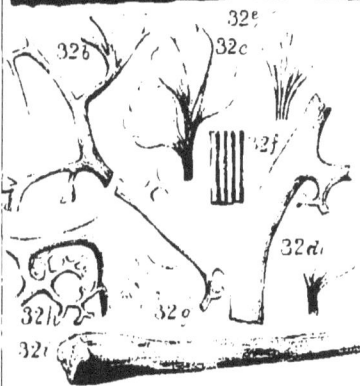

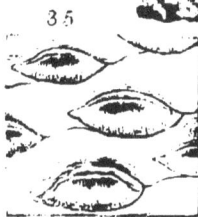

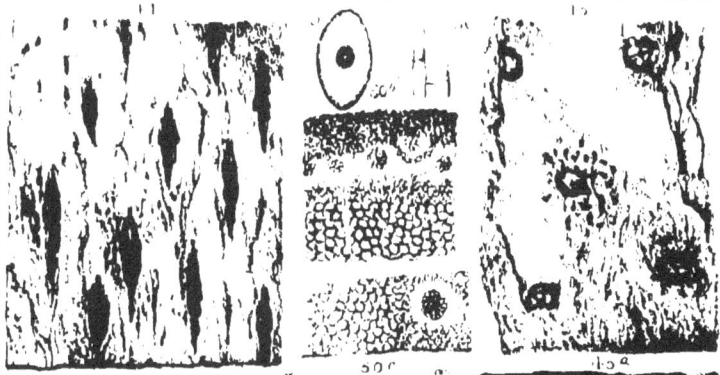

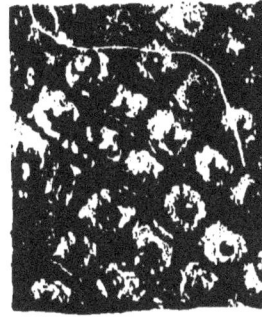

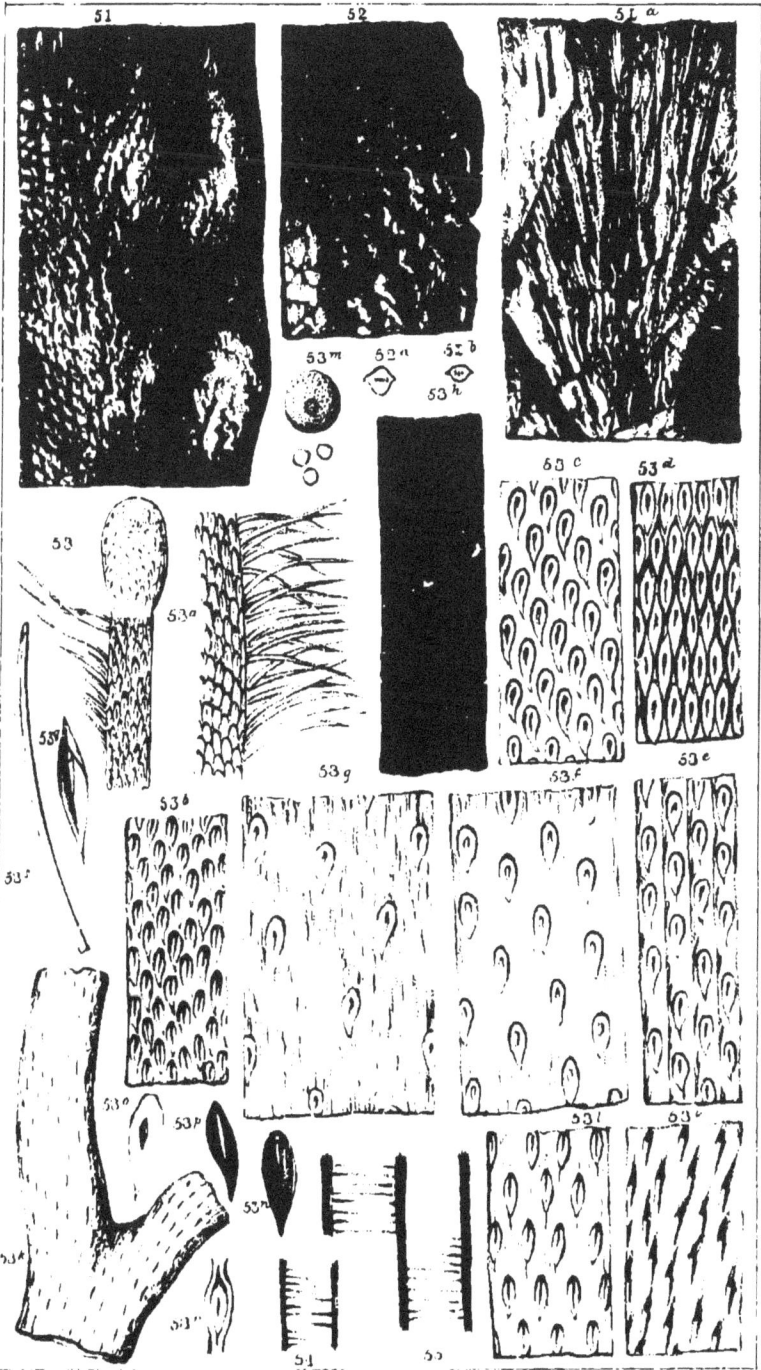

51   52   51 a

53 m   52 a   53 b
53 h

53   53 c   53 d

53 a   53 g   53 f   53 e

53 b

53 o   53 p

53 n   53 i   53 k

51   53

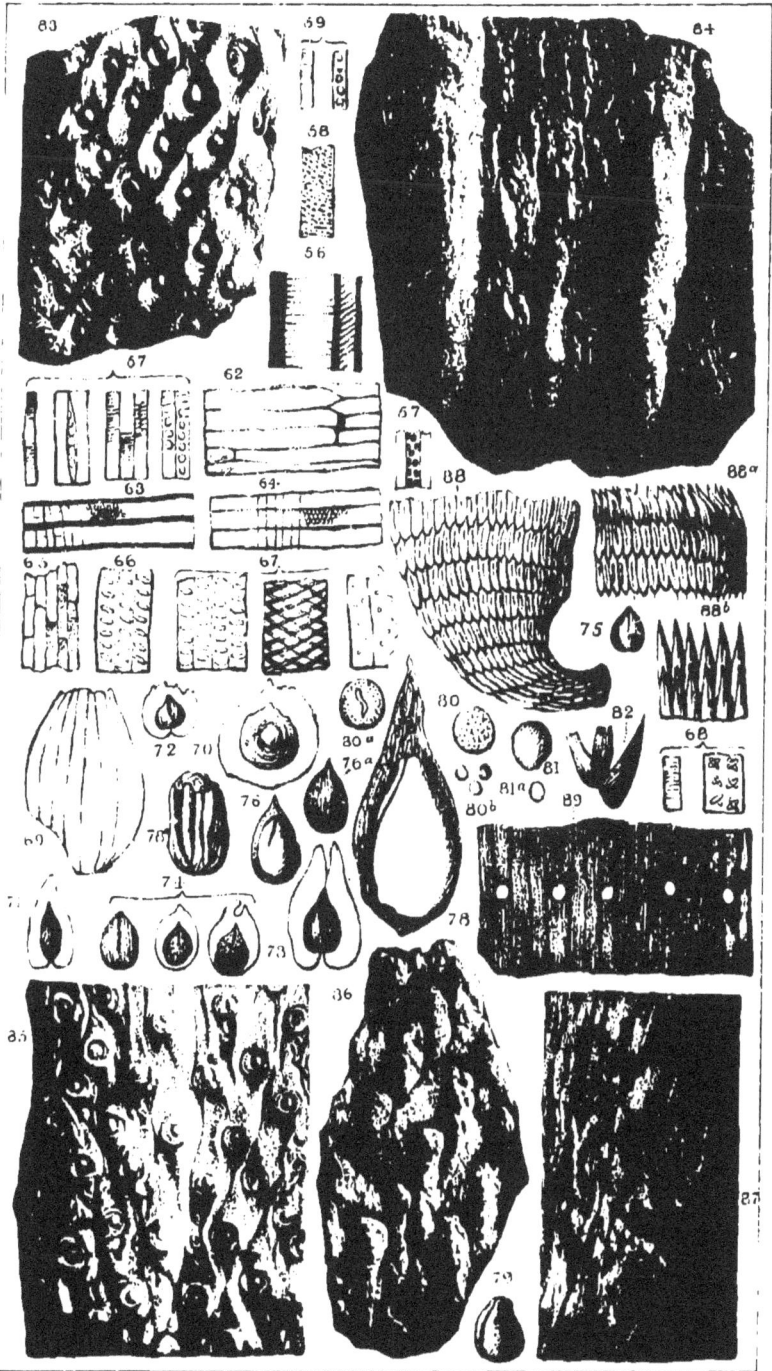